Measurement, Uncertainty and Lasers

Measurement, Uncertainty and Lasers

Masatoshi Kajita
National Institute of Information and Communications Technology, Tokyo, Japan

IOP Publishing, Bristol, UK

ISBN   978-0-7503-2328-4 (ebook)
ISBN   978-0-7503-2326-0 (print)
ISBN   978-0-7503-2327-7 (mobi)

DOI   10.1088/2053-2563/ab0373

Version: 20190401

IOP Expanding Physics
ISSN 2053-2563 (online)
ISSN 2054-7315 (print)

British Library Cataloguing-in-Publication Data: A catalogue record for this book is available from the British Library.

Published by IOP Publishing, wholly owned by The Institute of Physics, London

IOP Publishing, Temple Circus, Temple Way, Bristol, BS1 6HG, UK

US Office: IOP Publishing, Inc., 190 North Independence Mall West, Suite 601, Philadelphia, PA 19106, USA

I learned the fundamentals of laser spectroscopy from Profs T Shimizu when I was a graduate course student. After joining the Communications Research Laboratory (in 2004, since renamed the National Institute of Information and Communications Technology), I learned the fundamentals of the precise measurement of atomic transition frequencies from Prof. K Nakagiri, Prof. S Urabe, Dr. T Morikawa, and Dr. M Hosokawa. I recognized the important role of frequency precise measurement for all fields of physics after joining the meetings of 'Fundamental Physics Using Atoms', organized by Prof. N Sasao, Prof. K Asahi, Prof. Y Sakemi, Prof. K Sugiyama, and Dr. T Aoki.

At the publication of this book, I appreciate all the people listed above a lot.

Contents

Preface

The measurement of physical values is fundamental to physics. Note that all measurements are to a degree of uncertainty. Physics is a law of nature established on measurement results, which can be violated after the reduction of measurement uncertainties. The development of physics was closely correlated to the reduction of measurement uncertainties. The measurement uncertainty can create some serious problems, not only for science, but also for our daily lives.

In the last few decades, measurement uncertainties have been drastically reduced for time and frequency, to the order of 10^{-18}, which made it possible to also reduce the uncertainties of all other physical values. The role of lasers is very important for this revolution.

A laser is a light with a uniform phase, narrow frequency linewidth, and uniform propagating direction. Since the invention of the frequency comb, a laser frequency can be measured with an uncertainty of the order of 10^{-18}. With modern physics, the speed of light is absolutely constant. Therefore, the wavelength of laser light with an accurate frequency can be used as a good length scale. Using lasers, we can get the detailed energy structure of atoms and molecules, which give the information of elementary particles. This book introduces the important roles of lasers for the significant reduction of measurement uncertainties. There are still unsolved mysteries, and further reductions of measurement uncertainties are required to solve them.

I will feel happy if this book acts as a stimulant for young students to become interested in precise measurement with lasers. Also, to make it easier for students, I tried to compose my explanations using simple words without complicated equations.

Acknowledgments

The research activity of the author is supported by a Grant-in-Aid for Scientific Research (B) (Grant No. JP 17H02881), and a Grant-in-Aid for Scientific Research (C) (Grant Nos. JP 17K06483 and 16K05500) from the Japan Society for the Promotion of Science (JSPS). The author is highly appreciative of discussions with Y Yano, T Ido, N Ohtsubo, Y Li, H Hachisu, S Nagano, M Kumagai, and M Hara, all from the NICT, Japan, as well as T Aoki (the University of Tokyo), M Yasuda (AIST, Japan), and A Onae (AIST, Japan: passed away on 12 June, 2017). The author is grateful to K Kameta and J Navas (IOP, UK) for the opportunity to write this book.

Author biography

Masatoshi Kajita

Born and raised in Nagoya, Japan, Dr Kajita graduated with a degree in applied physics, from the University of Tokyo in 1981 and obtained his PhD in physics from the University of Tokyo in 1986. After working at the Institute for Molecular Science, he joined the Communications Research Laboratory (CRL) in 1989. In 2004, the CRL was renamed the National Institute of Information and Communications Technology (NICT). In 2009, he was a guest professor at the Provence Universite, Marseille, France.

IOP Publishing

Measurement, Uncertainty and Lasers

Masatoshi Kajita

Chapter 1

All measurements have uncertainties

1.1 Introduction

The measurement of values is a fundamental part of science and technology and strongly influences our real lives. We must recognize that all measurements involve a degree of uncertainty. When we make an agreement or contract related to a certain value, we consider the exact value as fair, but we accept small errors or variances. When we are invited to a party at someone's home starting at 19:00, it is considered acceptable to arrive at 18:55 or 19:05 (but not acceptable to arrive 18:00 or 20:00). We must make allowances for slight measurement errors because there are always measurement uncertainties.

Here, the tale of *The Merchant of Venice* by William Shakespeare illustrates the fact that measurement uncertainties are inescapable. The story is summarized as follows.

Antonio (merchant of Venice) had a debt from a money lender, Shylock. They had a contract that Antonio would give one pound of his flesh if he could not pay back the debt by the deadline. However, Antonio received news that his trade ships had met a storm and sank into the sea (it turned out to be a false story afterwards). Shylock appealed to the court to get Antonio's flesh. In court, Portia (the wife of Antonio's friend) impersonated a judge. At first Portia asked Shylock to give a mercy to Antonio, but Shylock refused and asked for justice. Portia permitted Shylock to take Antonio's flesh but said that it must be exactly one pound. If it were slightly more or less, Shylock would be put to death. Shylock was happy to hear this. However, Portia said 'This contract does not give you any blood. You must take Antonio's flesh without bleeding.' Of course, that would be impossible, so Shylock gave up getting Antonio's flesh.

In this tale, Shylock should have given up getting Antonio's flesh just with the requirement of 'exactly one pound'. An exact value is fundamentally impossible. There is a difference of the order of µg between different 1 kg mass standards. To make a strict contract, the volume of flesh should be for example 'between 0.99 and

1.01 pound'. Portia asked Shylock for mercy, but she expected he would refuse all mercy, and this was part of her strategy. Shylock could not get anything by simply asking for justice.

In shopping, we buy beef or pork with a price given for a mass (for example 100 g), and the mass is measured with a scale. There is a question whether the scale measures the mass accurately, but we accept its uncertainty. For a submission deadline, allowances can be made if submissions are handled by a person; nobody will refuse to receive a submission that is received just one minute after the deadline. However, there is no flexibility when submission deadlines are handled by a computer system. For the deadline of 12:00 controlled by a computer system, no submission will be received at 12:01. In this case, the deadline is given by a clock in the computer system, which has also some error. A submission should be completed sometime in advance of the deadline, considering that the deadline according to the computer system might come a little earlier than the real time.

This chapter shows how we have coexisted with non-zero measurement uncertainty. Section 1.2 briefly summarizes the history of the treatment of records in various sports. When the difference between records is smaller than the measurement uncertainty, we cannot determine victory or defeat. This is a good example of the treatment of measurement uncertainty. We must work to reduce measurement uncertainties because slight measurement errors can lead to serious problems in our lives, as discussed in section 1.3. There are also some phenomena that lead to significant differences in solutions with only slight differences in the initial state because of the nonlinearity of the equation to show temporal expansion. The uncertainty of the initial state cannot be exactly zero; therefore, we cannot evaluate the future with these phenomena (called 'chaos'), as explained in section 1.4.

1.2 Measurement uncertainty in sports

There are many sports that involve races, including athletics, swimming, speed skating, and so forth. However, at the time of the first modern Olympic Games in Athens in 1896, victory or defeat was determined only for simultaneous running [1]. Although there were time records, they were treated only as reference records because there was an uncertainty of 0.2 s with the stopwatches that were used at that time.

Since the Antwerp Olympics in 1920, stopwatches showing times with units of 0.01 s have been officially used, but results were still recorded with units of 0.2 s. This is because the measurement reliability was not accurate enough with hand measurement. For the Los Angeles Olympics in 1932, the results were recorded with units of 0.1 s. At the Helsinki Olympics in 1952 electric time measurement was used for the first time. Since the Munich Olympics in 1972, records were given with units of 0.01 s.

Now, measurement with units of 0.001 s is possible. In athletics the winner of a race is often determined with a difference smaller than 0.01 s. Why are records still given in units of 0.01 s? In athletics, we can distinguish the winner of a simultaneous running event from the difference of position at the finish line by several cm and if the time difference is less than 0.01 s (also by using photos). However, comparing the recorded times from different races, we cannot guarantee equal course lengths with a measurement uncertainty of 1 mm, as shown in figure 1.1. With swimming, races if the time difference is less than 0.01 s they are treated as tied (the course length might be different by mm in different lanes). At the Los Angeles Olympics in 1984, two swimmers received gold medals in the women's 100 m freestyle.

Winners are determined by time records with units of 0.01 s for Alpine skiing, sled competitions, speed skating and so forth. Only for the 500 m speed skating are ties broken with a difference of 0.001 s. Competitors skate the same course one by one (speed skating by two); therefore, the equality of the course length is guaranteed. However, the equality of the conditions (course surface, wind, etc) is questionable, and it is not useful to compare time records with differences smaller than 0.01 s. In women's downhill skiing at the Sochi Olympics, two skiers received gold medals having tied records. In ski jumping, the flight distance is measured with an accuracy of 0.5 m, because a difference smaller than 0.5 m gives a score difference (2.0/m for the normal hill and 1.8/m for the large hill) that is negligible in comparison with the score difference for the jumping style, which is subjectively determined by referees.

The course length of a marathon is set to be 42.195 km, but it has a measurement uncertainty. It is regulated that the course length measured at 30 cm from the edge (kerb) of the course (for curves, the shortest length) should be between 42.195 and 42.2372 km. This regulation accepts an uncertainty of 0.1%, but the course length must not be shorter than 42.195 km even by 1 mm. In the New York City marathon in 1981, A Salazar and A Roe set the world records for men's and women's marathons, respectively [2]. However, both records were canceled in 1984 because it was realized that the course was 148 m too short (at that time, both records had already been updated). Marathon time records are given with in units of 0.1 s because the course length cannot be measured accurately enough to consider a difference of 0.01 s. Moreover, it is useless to compare the time records of different races with different slopes and weather.

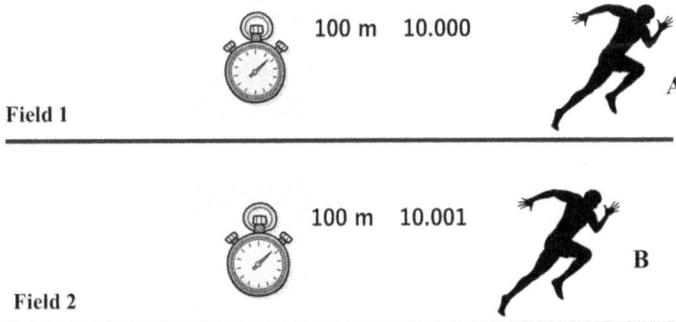

Are both course lengths exactly same?
Is runner A really faster than runner B?

Figure 1.1. Comparison of 100 m races in different places.

1.3 Are uncertainties always acceptable?

1.3.1 DNA identification in criminal trials

As previously mentioned, all measurements have some uncertainties, but we must recognize that this uncertainty can lead to serious problems. For example, criminal trials are sometimes based on DNA identification. With the DNA identification method that was available in the 1990s, the result could match to that of another person with an uncertainty of 0.01% (figure 1.2). In 1990, a four year-old girl was murdered in Ashikaga, Japan. The accused was convicted and sent to prison based on the result of DNA identification as evidence of guilt [3]. In 2009, DNA identification was performed again by both the prosecution and defense. The accuracy of DNA identification had improved greatly during that time, and the possibility of incorrect DNA matching was less than 10^{-12}. Both results proved the convict's innocence and he was released. In this example, measurement uncertainty led to an innocent person being sent to prison. Clearly, we should not trust measurement results perfectly.

1.3.2 Position measurement on the Earth

When we are in an unfamiliar place, we may lose our way. When this happens determining our location can be easier when we see a tall landmark that is visible from a distance (for example the Eiffel Tower in Paris). This is helpful on land, but what can we do in the ocean, where there are no landmarks? In this case, we can find our position by observing the stars. In the Age of Discovery (15th–18th centuries), navigators found it necessary to measure their positions. Latitude was measured from the height of Polaris above the horizon. Longitude was measured by the position of the constellations at a given time of day; therefore, the role of the clock was very important. Position measurement was difficult when the stars could not be seen, for example in cloudy weather. In such situations it was possible for navigators to lose their way.

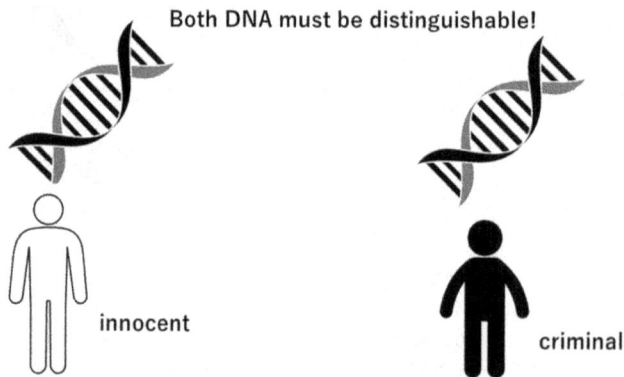

Figure 1.2. DNA identification as evidence for a criminal trial.

We can lose our position in a small area on a mountain, for example, in a snowstorm. In 1902, a group of 210 travelers who were going through Hakkouda Mountain (Japan) met a severe snowstorm, and 199 of them died [4].

A mistake regarding the flight route of Korean Air 007 led to a tragedy when the plane entered Russian airspace and was shot down (the cause of this mistake is still a mystery). Since then, the use of the global positioning system (GPS, developed by US military in 1973) [5] has been available for civil use to avoid flight course errors. The fundamental function of GPS is to measure the distance between satellites and receivers from the propagation time of radio waves. Measuring the distances with three satellites, the position of the receiver (latitude, longitude, and altitude) can be determined as shown in figure 1.3. The propagation of radio waves is not disturbed by clouds, and the receiver can always receive radio waves from more than six satellites. Therefore, position measurement by GPS is possible at any time in any weather.

The clocks in the satellites are highly accurate atomic clocks giving an error of only 10^{-6} s which results in a distance error of 300 m. Although the clock in the receiver is a quartz clock, its error is corrected by receiving radio waves from more than four satellites. For the US military to maintain their advantage over foreign military forces, intentional degradation of accuracy, called selective availability (SA), was added to civil users' times. While SA was operating, the position accuracy was on the order of 100 m. In 2000, the US government discontinued the use of SA. The main reason was that other countries were also launching their own satellites, and the US no longer remained the sole provider of space-based position measurements. Now, an accuracy of a few meters can be achieved with civilian receivers.

The radio waves from satellites cannot be received in a submarine cruising at an abyssal. In this case, an inertial navigation system is used [6]. This system measures the acceleration (using an accelerometer) and rotation (using gyroscopes). The motion direction against the Earth's magnetic field is also monitored using magnetometers. The velocity is obtained by integrating the acceleration and the position is estimated by integrating the velocity. The uncertainty of the position becomes larger as time goes

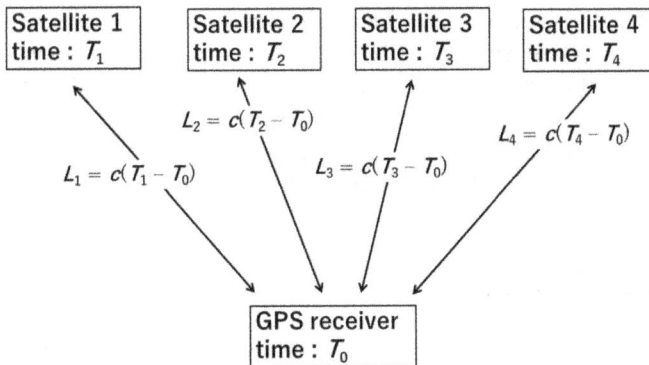

Figure 1.3. Position measurement by GPS. Distances from four satellites are calculated from the propagation time of the electromagnetic wave.

by due to the integration of small errors. Therefore, periodic correction by another method (for example GPS measurement) is also required. Inertial navigation is also used for ships, aircraft, guided missiles, and spacecraft.

1.3.3 Position measurement in space

With a slight uncertainty of its position and velocity, a spacecraft cannot arrive at its intended destination. Cassini–Huygens (launched in 1997) entered the orbit of Saturn in 2004 after flying past Venus (1998–99), Earth (1999), and Jupiter (2000) [7]. This original mission was to probe Saturn from 2004 to 2008, but it continued until 2017. The motion of such spacecraft must be controlled with very high accuracy because the revolution motion velocities of planets are very fast (see table 1.1).

With the range and range rate (RARR) method (see figure 1.4), the position of a probe is monitored by the propagation time of radio waves irradiated and reflected from the spacecraft [8]. The velocity is monitored from the shift of the frequency of the reflected radio wave given by the Doppler effect (shift of the frequency of waves from moving objects). The position and velocity accuracies are a few meters and a few millimeters per second, respectively, in the propagation direction. However, the accuracy in the direction perpendicular to the motion cannot be high. The accuracy of direction is 10^{-6} rad observing with one observatory (NASA used three observatories, and this accuracy was 2.5×10^{-7} rad). Then the position uncertainty with the distance of 250 million km is 250 km, which is not accurate enough for planetary exploration.

Very-long-baseline interferometry (VLBI) is used to improve the accuracy of the motion direction. This method has been used to measure the distance between two observatories from the slight difference in the detection times of a pulse signal from a quasar, whose direction is known in detail. It is also possible to determine the direction of a spacecraft from the difference in detection times of a signal from a spacecraft. By detecting the signals from a quasar and an aircraft simultaneously (called differential VLBI), we can measure the motion direction with higher accuracy because of the significant cancellation of noise on both signals caused by the air and ionosphere. With the combination of RARR and differential VLBI, the position uncertainty is reduced by one order more than using just RARR [8].

Recently, the Japanese spacecraft Hayabusa 2 arrived at the asteroid Ryugu. The size of Ryugu is 900 m, and it is 300 million km away from the Earth. The required accuracy of the flight course is equivalent of hitting an object of 6 cm from a distance of 20 000 km (the distance between the UK and New Zealand), which is not possible with the method described above. First of all, the position of each asteroid is not

Table 1.1. Revolution motion velocities of planets.

Moon 1.01 km s^{-1}			
Mercury 47.9 km s^{-1}	Venus 35.0 km s^{-1}	Earth 29.4 km s^{-1}	Mars 24.1 km s^{-1}
Jupiter 13.1 km s^{-1}	Saturn 9.7 km s^{-1}	Uranus 6.8 km s^{-1}	Neptune 5.5 km s^{-1}
Sun 220 km s^{-1}			

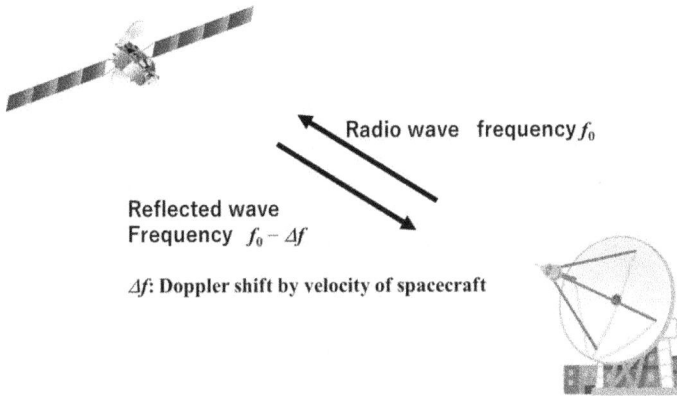

Figure 1.4. Position and velocity measurement by the range and range rate (RARR) method.

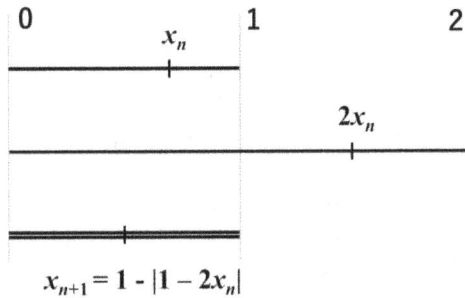

Figure 1.5. Baker's transformation.

known in detail. Therefore, Hayabusa 2 has a system to monitor the position of Ryugu using an onboard camera (optical navigation) and correct its orbit automatically [9, 10].

1.4 Chaos

Measurement uncertainty can lead to situations characterized by chaos in which we cannot estimate future results by solving equations. This is induced by the non-linearity of the equation, in which a slight difference in the initial value leads to an exponential growth of difference with time evolution. As the simplest example to illustrate chaos, we present a repetition of Baker's transformation below (see figure 1.5) [11, 12].

(1) We consider the position of x_n on a bar, whose ends are 0 and 1.
(2) The bar is expanded to double its original length and then collapsed to half the original length. Then x_n is transformed to $x_{n+1} = 1 - |1 - 2x_n|$.

Figure 1.6 shows x_n taking $x_0 = 0.3$ and 0.300 003. With $n > 15$, both the differences of x_n become significant. As x_0 must have some uncertainty, the value of x_n becomes unpredictable after repetition of this transformation.

Baker's transform

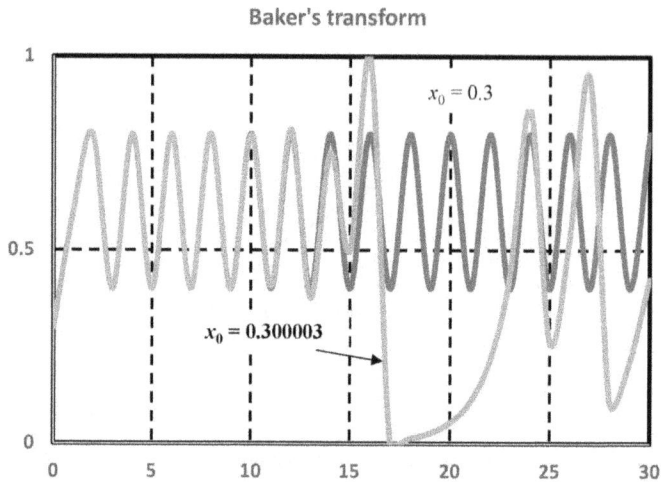

Figure 1.6. Baker's transformation with the initial values of 0.3 and 0.300 003.

The concept of Baker's transformation shows that the chaos phenomenon is observed as a function of time $x(t, x_0)$ (x_0: initial value of x), satisfying $x_{min} < x(t, x_0) < x_{max}$ and $|(x(t, x_0 + \delta x) - x(t, x_0))/x(t, x_0)| = |\delta x/x_0|e^{(t/t_0)}$ ($t_0 > 0$). The value of $T_0(= t_0 \ln[|x_{max} - x_{min}|/|\delta x|])$ is a parameter that we can use to predict the value of $x(t)$ only with $t < T_0$.

The concept of chaos was introduced by Lorenz as the Lorenz's equation to predict the weather [13]. Although an equation is established, its solution a month later becomes quite different with a slight difference in the initial condition. Weather is predictable for the next week, but prediction for a longer time-scale is not realistic.

References

[1] http://olympstats.com/2014/02/12/timing-accuracy-at-the-olympic-games/
[2] http://runblogrun.com/2011/10/alberto-salazars-wb-marathon-102581-thirty-years-and-three-days-by-jeff-benjamin-note-by-larry-eder.html
[3] https://yomiuri.co.jp/adv/chuo/dy/opinion/20100419.html
[4] https://en.wikipedia.org/wiki/Hakk%C5%8Dda_Mountains_incident
[5] https://faa.gov/about/office_org/headquarters_offices/ato/service_units/techops/navservices/gnss/gps/
[6] https://en.wikipedia.org/wiki/Inertial_navigation_system
[7] https://en.wikipedia.org/wiki/Cassini%E2%80%93Huygens
[8] Yoshikawa M and Nishimura T 2000 計測と制御 (in Japanese) **39** 564
[9] http://hayabusa2.jaxa.jp/en/topics/20180828eb/index.html
[10] http://hayabusa2.jaxa.jp/topics/20180515_e2/
[11] https://de.wikipedia.org/wiki/B%C3%A4cker-Transformation (in German)
[12] Inoue M 1997 「やさしくわかるカオスと複雑系科学(in Japanese)」日本実業出版社 ISBN4-53402492-4
[13] Lorenz E N 1963 *J. Atmos. Sci.* **20** 130

Chapter 2

What is measurement uncertainty?

2.1 Introduction

To reduce measurement uncertainties, we must know their properties. In this chapter, we present various kinds of measurement uncertainties and methods to reduce them. We also introduce parameters to estimate measurement uncertainties, which depend on the use of measurements.

2.2 Statistical and systematic uncertainties

When we measure something, there is a question whether it is a real value. The reliability of measured values can be confirmed by repeating measurements many times. We will see that measurement results are distributed over a certain limited area. The uncertainty given by the finite distribution area is called 'statistical uncertainty' as shown in figure 2.1.

We cannot guarantee that the real value is in the area over which the measurement results are distributed. All measurement values can vary according to circumstances, and the real values should be defined with a certain condition. If we take measurements under another circumstance, the measured value may shift from the defined value. If we know the dependence of the measurements on the circumstances, we can make a correction of the measured values by the estimated shift. Then the uncertainty of the estimated shift becomes another uncertainty, called 'systematic uncertainty' as shown in figure 2.1.

More detailed explanations of statistical and systematic uncertainties are presented in the following subsections.

2.2.1 Statistical uncertainty

Statistical uncertainty is given by the finite broadening of the distribution area of the measurement results. This broadening can be induced by the temporal fluctuation of

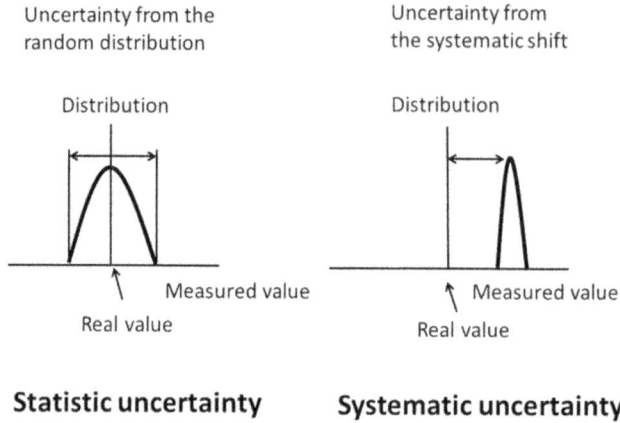

Statistic uncertainty **Systematic uncertainty**

Figure 2.1. Statistic and systematic uncertainties with the distributions of measurements.

the circumstances, which can be reduced by stabilizing the circumstances. This broadening is also induced by the quantum uncertainty principle.

We consider the measurement of a physical value X. The measurement results are distributed around the real value with the measuring condition X_r. Taking measurement samples $X(i)$ ($i = 1 - N$), we obtain the average X_{ave} and the standard deviation σ. When N is large enough, the probability distribution of X_{ave} is approximately given by (figure 2.2)

$$P(X_{\text{ave}}) = \frac{\sqrt{N}}{\sigma\sqrt{\pi}} \exp\left[-N\left(\frac{X_{\text{ave}} - X_r}{\sigma}\right)^2\right] \tag{2.1}$$

with any distribution of the measurement results (central limit theorem). Therefore, X_r is estimated to be in the range of

$$X_r = X_{\text{ave}} \pm \frac{\sigma}{\sqrt{N}}. \tag{2.2}$$

The statistic uncertainty is reduced by increasing the number of measurement samples.

It is not simple to derive the central limit theorem as a general formula [1]. Here, we derive it using the simplest model by which we get the measurement results $X_r + \sigma$ and $X_r - \sigma$ with the probability of 1/2. Repeating measurement N times, the probability $p(n)$ to measure n times $X_0 + \sigma$ and $(N - n)$ times $X_0 + \sigma$ is given by

$$p(n) = \frac{N!}{n!(N - n)!}\left(\frac{1}{2}\right)^N. \tag{2.3}$$

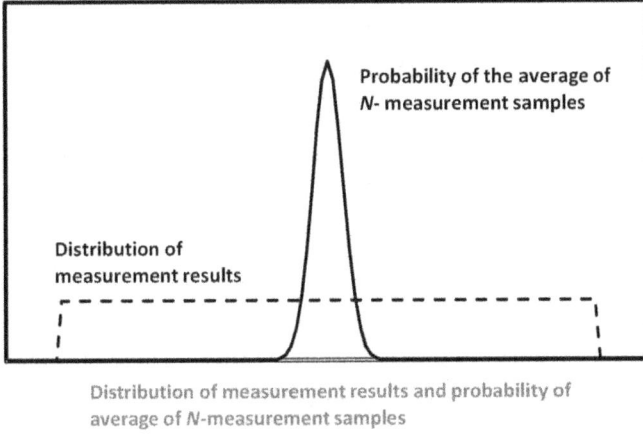

Distribution of measurement results and probability of
average of N-measurement samples

Figure 2.2. Relation between the distributions of measurement results and the probability of the average of
N-measurement samples. The probability of the average of the N-measurement samples is given by a Gaussian
with broadening narrower than that of measurement results with a factor of $1/\sqrt{N}$.

Here, we consider the Taylor expansion of $\ln[p(n)]$ using

$$\frac{d \ln[p(n)]}{dn} = -\ln(n) + \ln(N - n),$$

$$\frac{d^2 \ln[p(n)]}{dn^2} = -\frac{1}{n} - \frac{1}{N - n}. \tag{2.4}$$

Here, $\ln[p(n)]$ at n close to $N/2$ is approximately given by

$$\ln[p(n)] = \ln p\left(\frac{N}{2}\right) - \frac{4}{N}\left(n - \frac{N}{2}\right)^2,$$

$$p(n) = p(0)\exp\left[-\frac{4}{N}\left(n - \frac{N}{2}\right)^2\right]. \tag{2.5}$$

The average of the measurement is given by

$$X_{\text{ave}} = X_r + \frac{2n - N}{N}\sigma,$$

$$n - \frac{N}{2} = \frac{N}{2\sigma}(X_{\text{ave}} - X_r). \tag{2.6}$$

Then the probability of the average of N-sample measurements is given by

$$P(X_{\text{ave}}) = P(X_r)\exp\left[-N\left(\frac{X_{\text{ave}} - X_r}{\sigma}\right)^2\right]. \tag{2.7}$$

2.2.2 Systematic uncertainty

Statistical uncertainty is reduced by averaging many measurement samples. However, it is possible that all measurement results have a parallel shift from the real value. This is because all the measured values depend on the circumstance under which they were taken. It is not always possible to measure the values while measuring the defined value. Seeing just the measurement results, the shift of the measured value is the systematic uncertainty. Systematic uncertainty is reduced by monitoring the circumstance and giving a correction of the estimated shifts. With this correction, the systematic uncertainty is given by the uncertainty of the estimated measurement shift.

For example, thermal expansion causes a shift in the length of an object; therefore, at which temperature (T_0) the length is defined should be clarified. Repeating the measurement with another temperature T_p, the averaged length is shifted from the defined length (figure 2.3). The systematic uncertainty is reduced by monitoring T_p and giving a correction by $\alpha_p(T_0 - T_p)$, but it cannot be zero because of the uncertainties of T_p and α_p.

2.2.3 Uncertainty in atomic transition frequencies

Measurement uncertainty is particularly low with the frequency of electromagnetic waves absorbed or emitted by atoms or molecules in the gaseous state (atomic, molecular transition frequencies). Using this characteristic, atomic clocks based on the atomic transition frequencies were invented after World War II, and the uncertainty of time and frequency was reduced drastically (chapter 3). However, atomic transition frequencies also have non-zero measurement uncertainties. Here, we discuss the factors that contribute to the measurement uncertainties of atomic transition frequencies.

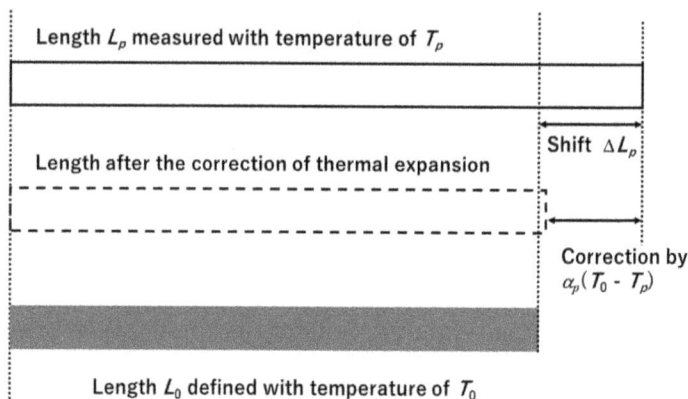

Figure 2.3. Concept of systematic uncertainty with length measurement. The length should be defined with a certain temperature T_0. Measuring with another temperature T_p, there is a shift of measurement and this gives a systematic uncertainty. This uncertainty is reduced by monitoring the temperature and giving a correction of thermal expansion $\alpha_p(T_0 - T_p)$ (α_p: thermal expansion coefficient), but it cannot be zero because of the uncertainties of T_p and α_p.

Statistical uncertainty is given by the limited interaction time τ_e between atoms and electromagnetic waves without the phase jump. The limit of τ_e is caused by several things: limited interaction time between atom and electromagnetic wave by the atomic motion, collision between atoms, or spontaneous emission transition (transition from a higher energy state to a lower energy state giving fluorescence). The transition frequency has the uncertainty of the order of $\delta f = 1/2\pi\tau_e$, because the phase procedure $2\pi f \tau_e$ must have the uncertainty of ± 1. This relation is derived from the quantum uncertainty principle between time and energy. With the real transition frequency of f_0, the distribution of the single measurement result is given by

$$I(f) = \frac{\delta f}{(f - f_0)^2 + (\delta f)^2},$$

(2.8)

and δf is observed as the spectrum linewidth. The statistical uncertainty δf_{sta} is given by

$$\delta f_{\text{sta}} = \delta f \sqrt{\frac{(\tau_e + \tau_d)}{N_a T_m}} = \frac{1}{2\pi\tau_e} \sqrt{\frac{(\tau_e + \tau_d)}{N_a T_m}},$$

(2.9)

where N_a is the number of atoms, τ_d is the measurement dead time (time for preparation or detection), and T_m is the measurement time.

The systematic measurement uncertainty is given by the shift of the measured atomic transition frequency caused by various things. For example, the electron orbit in the atom is distorted by the electric or magnetic fields, and the transition frequencies are shifted. The frequency shift caused by the electric field and the magnetic field are called Stark and Zeeman shifts, respectively. The Stark shift cannot be zero because the electric field is applied by the electromagnetic wave for the probe and by blackbody radiation (radiation of electromagnetic wave from objects having non-zero temperature). The Zeeman shift also exists because of the Earth's magnetic field.

Systematic uncertainty is also given by the relativistic effects as follows (chapter 6). The motion of atoms (at room temperature on the order of 200–500 m s^{-1}) gives frequency shifts called quadratic Doppler shifts. The difference in the gravitational potential also gives the frequency shift called the gravitational red shift (effect of the theory of general relativity).

The definition for the atomic transition frequency is given by the condition of zero velocity free from electric and magnetic fields on the geoid plane. Atomic clocks are based on atomic transition frequencies, where the statistical and systematic uncertainties are relatively small [2–6]. Atomic laser cooling (chapter 4) made it possible to reduce the atomic velocity to a few cm s^{-1}, and the quadratic Doppler shift was reduced significantly. Statistic uncertainty was also reduced since the development of laser cooling because of the longer interaction time between atoms and electromagnetic waves. Uncertainty on the order of 10^{-18} was obtained with several atomic transition frequencies after correction of the possible frequency shifts.

2.3 Accuracy and stability

The measurement uncertainty is discussed with regard to accuracy and stability. Figure 2.4 shows the distributions of measurements with regard to 'high' and 'low' levels of 'accuracy' and 'stability'. Accuracy indicates the reliability of the final results after the averaging of many experimental results and correction of measurement shifts induced by various causes. Accuracy can also be high if the difference between each measurement result is large. To achieve high accuracy, the systematic uncertainty should be small because the statistical uncertainty can be reduced by repeating a measurement many times. The accuracy is estimated from the possible shift of measurements by theoretical calculation or experimental measurement with different circumstances. Comparison of measurements obtained using different devices is also performed.

Stability refers to the constancy of each measurement result. The stability can be high if the measurement results are equally shifted from the real value. During the repetition of a measurement over a long period, there may be temporal change of the measurement values induced by changes in the circumstance. Therefore, we distinguish 'short-term stability' and 'long-term stability' from the averaging time. Short-term stability is determined from the statistical uncertainty covering short measurement times, with which the change of the circumstance is negligibly small. To achieve high short-term stability with the atomic transition frequency, a narrow spectrum linewidth (δf in equation (2.9)) and a large number of atoms (N_a in equation (2.9)) are required. With long measurement times, the statistical uncertainty is suppressed, but the influence of the change in conditions (e.g. electric field and magnetic field) becomes significant. Therefore, long-term stability is determined by the systematic uncertainty. The standard deviation is generally used for the

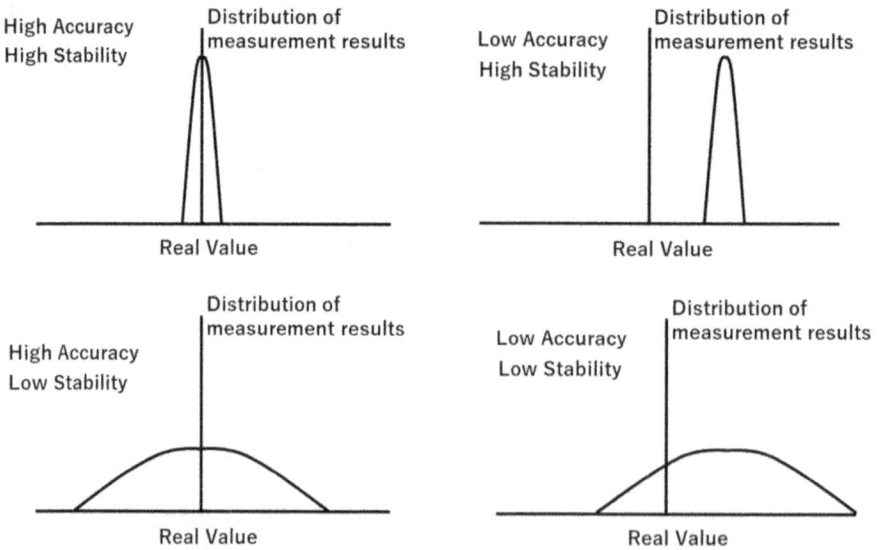

Figure 2.4. Schematic of the distribution of measurements with high and low 'accuracy' and 'stability'.

estimation of statistical uncertainty, but it is not useful to include measurements with a temporal change of circumstance. With the linear drift of measurement value, the standard deviation is divergent.

Particularly for the estimation of frequency stability, Allan variance is often used for any averaging time [7]. The Allan variance is given by

$$\sigma_y(M, T_a, \tau) =$$
$$\sqrt{\frac{1}{M-1}\left\{\sum_{i=0}^{M-1}\left[\frac{x(iT_a + \tau) - x(iT_a)}{\tau}\right]^2 - \frac{1}{M}\left[\sum_{i=0}^{M-1}\frac{x(iT_a + \tau) - x(iT_a)}{\tau}\right]^2\right\}}, \qquad (2.10)$$

where $x(t)$ denotes the phase at the time t. Here, $|x(iT_a + \tau) - x(iT_a)|/\tau$ shows the average frequency between iT_a and $iT_a + \tau$. The Allan variance is the variation of

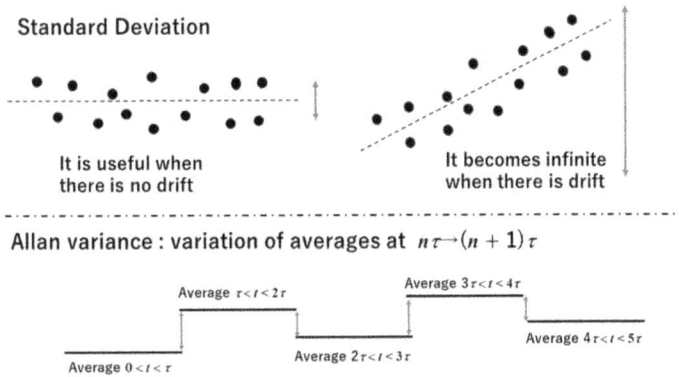

Figure 2.5. Concepts of standard deviation and Allan variance with $M = 2$ and $T_a = \tau$ (equation (2.7)). Standard deviation is not useful when there is a drift. Allan variance is the variation of the average at various measurement time periods, which is also useful when there is a drift.

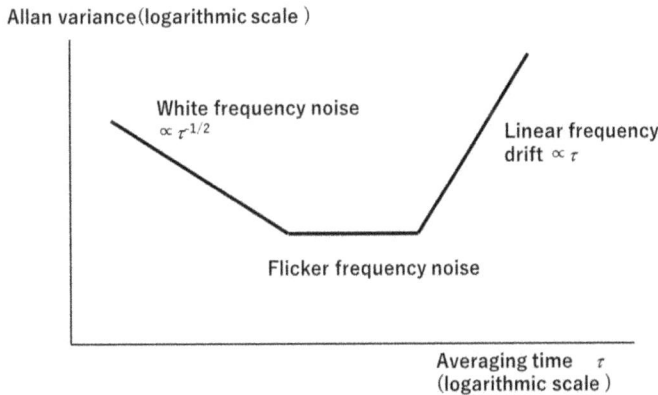

Figure 2.6. Dependence of the Allan variance on the averaging time τ considering the white frequency noise, flicker frequency noise, and linear frequency drift.

the averaged frequency for the period of τ at various time periods. Figure 2.5 shows the case with $M = 2$ and $T_a = \tau$. Figure 2.6 shows the dependence of the Allan variance on τ.

The short-term stability is generally given by the white frequency noise, and the Allan variance is proportional to $1/\sqrt{\tau}$, as shown with the standard deviation. With longer τ, the flicker frequency noise (called $1/f$ noise) is more significant than the white frequency noise, and the Allan variance is constant with τ. When there is a linear frequency drift, the Allan variance is proportional to τ with large τ.

References

[1] https://en.wikipedia.org/wiki/Central_limit_theorem
[2] Heavner T P *et al* 2005 *Metrologia* **42** 411
[3] Clairon A, Salomon C, Guellati S and Phillips W 1991 *Europhys. Lett.* **16** 165
[4] Chou C W *et al* 2010 *Phys. Rev. Lett.* **104** 070802
[5] Ushijima I *et al* 2015 *Nat. Photon.* **9** 185
[6] Nicholson T L *et al* 2015 *Nat. Commun.* **6** 6896
[7] https://en.wikipedia.org/wiki/Allan_variance

Chapter 3

Units of physical values and their definitions until 1960

3.1 Introduction

To reduce measurement uncertainty, unified definitions of fundamental physical values are required. For example, length measurement using different scales produces different values. Therefore, we need to know how the exact length of 1 m is defined as a universal value. The defined standard values also have some uncertainties, but they must be much smaller than conventional measurement uncertainties. After measurement uncertainties are reduced, the previous definitions cannot be kept as standards. Therefore, new definitions of standard values are required. In this way, the definitions of standard values have been changed several times.

After the invention of lasers (chapter 4) in 1960, the measurement uncertainties of various physical values were drastically reduced. This chapter introduces the definitions of physical values (time and frequency, length, mass, temperature, electric current, luminal intensity, and substance quantity) before the invention of lasers.

3.2 Unification of units of physical values

The concepts of length and mass have been important for human lives since ancient times because they were needed to compare the values of goods to be exchanged. The length and mass of goods were measured with standard rulers or scales in trade. Although the concept of time was less strictly defined for our ancestors, time was measured using sundials or water clocks. These physical values have been expressed using different units in different geographic regions. Therefore, it was difficult to understand the physical values expressed in the units used in other regions. Even now, it can be inconvenient to buy clothing or shoes from suppliers in other countries that use different sizing systems. However, in ancient times there was no serious problem in this regard, because most people only moved and conducted business within a small area. Also, there was no necessity to understand foreign languages.

Since the age of discovery, people in different areas began to actively trade with one another. This led to inconvenience because sellers and buyers were using different units. The movement towards the unification of units started in France after the French revolution [1]. In 1790, Talleyrand-Perigord proposed the global unification of the unit of length. In 1791, the French Academy of Science advised adopting a definition of the unit of length, called the meter, to be one ten-millionth of the distance between the North Pole and the equator. The mass of 1 kg was also defined a that of a cube of water whose dimension was a decimal fraction of the unit of length. The Celsius temperature scale was established defining the freezing and boiling temperatures of water to be 0 °C and 100 °C, respectively. There was debate about the unification of units in many other countries, but it was not simple to switch from familiar units to unified units. At the universal Exposition in Paris in 1867, a group of scientists made a resolution about the unification of units based on the metric system. In 1875, the metric convention was signed by 17 states, aiming at the global unification of units. By the 1960s, the metric system was accepted by most countries.

However, the metric system is not used in everyday life all over the world. In everyday life in the US, the Imperial system is used. In the US, temperature is generally measured using the Fahrenheit scale. However, this is not a serious problem because Americans can also use the metric system in international contexts. Similarly, English is recognized as the global language, while cultures throughout the world keep their own mother languages. A global standard is required, while respect for the local culture is also important. The frequency of radio waves for television or radio broadcasting varies according to the area, but it must be standardized for communication between aircraft and control towers. The side of the road on which cars drive (left- or right-hand traffic) varies among countries, but the pedal placement is globally standardized to be clutch, break, and accelerator from left to right.

Physical phenomena are described using the International System of Units (SI) based on the following seven fundamental units:

Mass	kilogram (kg)
Length	meter (m)
Time	second (s)
Thermo-dynamical temperature	Kelvin (K)
Electric current	ampere (A)
Luminous intensity	candela (cd)
Substance quantity	mol

Several other physical values are given by the following:

Velocity	$m\ s^{-1}$	Acceleration	$m\ s^{-2}$
Force	Newton (N) = $kg\ m\ s^{-2}$	Energy	Joule (J) = $kg\ m^2\ s^{-2}$.

With the development of measurement technology, the uncertainty of the definitions of the units of physical values (for example a mass of 1 kg) became the limiting factor for the attainable measurement accuracy. Therefore, the unit values have been updated by new definitions with lower uncertainty. The precise measurement of time and frequency was much more difficult than that of length and mass in ancient times. Now the highest accuracy is obtained with time and frequency. Through the definition of several fundamental constants as values without uncertainties, all physical values are expected to be measured with the accuracy of time and frequency.

3.3 The standard of time and frequency (until Cs atomic clock)

The concept of time in ancient times was much less strictly defined than it is now. Around 3000 BC, sundials were used in Egypt to tell the time of day between sunrise and sunset [2]. A sundial consisted of a flat plate (dial) and a gnomon, which casts a shadow onto the dial (figure 3.1). Sundials cannot be used at night or during periods of cloudy weather. Around 1600 BC in Egypt and Babylon, water clocks were also used. There is also a claim that water clocks already existed in China around 4000 BC. Time was measured by the regulated flow of water into (inflow type) or out of (outflow type) a vessel (figure 3.1). The viscosity of water strongly depends on the temperature [3]; therefore, the flow rate was not constant. There was uncertainty of one or two hours per day in ancient times, but it did not cause any serious problem in human life.

Mechanical clocks were developed in medieval Europe. The first system involved gears that were rotated by an attached weight falling under gravity. They were

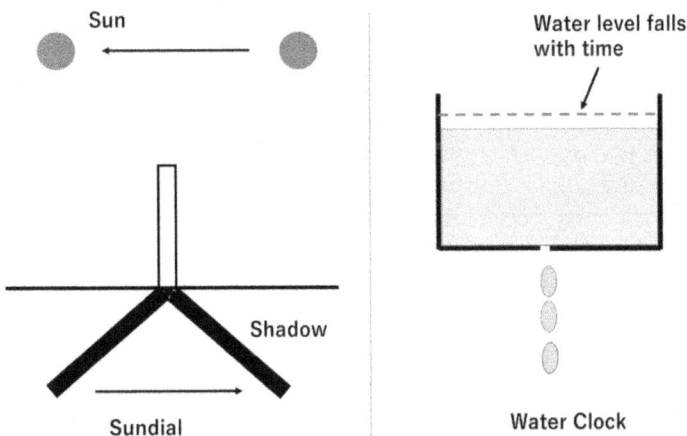

Figure 3.1. Clocks in ancient times: sundial and water clock.

placed in the high towers of churches to make the distance that the weight fell longer. They were used to announce the times for prayers with the chiming of bells [4]. There was only one clock hand on the mechanical clocks at that time because there was an uncertainty of the time of day on the order of one hour, and it was impossible to see a time difference on the order of several minutes.

In 1581, Galileo discovered the periodicity of the pendulum's swing, which enabled rapid improvements in the accuracy of clocks. In 1656, Huygens invented the pendulum clock. Initially, the error in the pendulum clock was 10 minutes per day, but later it was reduced to 1 minute per day with better gear-train mechanisms. Since then clocks have had two hands. Pendulum clocks were the most accurate time keepers until the 1930s [5]. Here, we consider the attainable uncertainty of time with a pendulum clock, which is based on the oscillation period of the pendulum expressed as

$$T = 2\pi\sqrt{\frac{l}{g}}, \tag{3.1}$$

where l and g are the length of the pendulum string and the gravitational acceleration, respectively. There is an uncertainty of time given by pendulum clocks; l is changed by thermal expansion, and g depends on the place.

A pendulum clock is not useful on a ship because g fluctuates significantly due to vibration. The British government offered a huge prize for the development of an accurate clock for use on a ship (as seen in section 1.3). In 1728, the first marine chronometer watch was developed, in which the weight is vibrated by a spring. Then the time measurement is independent of the gravitational force and can be used in moving ships. Four generations of marine chronometer watches were developed over the next 40 years gradually reducing the size, finally leading to the development of a compact pocket watch.

While new marine chronometer watches were developed, pendulum clocks were still useful for time measurement on land. In 1721, a mercury pendulum clock was developed that used a cylinder with a length of L_c filled with mercury at the bottom with a height of L_m as the pendulum, as shown in figure 3.2 [6]. The oscillation period is given by the length from the upper end of the cylinder to the center of mass of the mercury ($L_M = L_c - L_m/2$). With the change in temperature $T \to T + \Delta T$, there is a thermal expansion of the length of cylinder by $L_c \to L_c + \Delta L_c$ and that of mercury by $L_m \to L_m + \Delta L_m$, respectively. The values of L_c and L_m can be chosen so that $\Delta L_c = \Delta L_m/2$ is satisfied. Then L_M is constant when the temperature fluctuates. The mercury pendulum clock was the most accurate timekeeping device until the beginning of the 20th century.

The first quartz (SiO_2) clock was built in 1927, the accuracy of which was more than one order higher than that of pendulum clocks. Quartz is a piezoelectric material; that is, when a quartz crystal is subject to mechanical stress (such as bending), it accumulates electric charge across some planes. On the other hand, a quartz crystal will bend when electric charge is accumulated across the crystal plane. The vibrational motion of a quartz crystal is stimulated by the application of electric

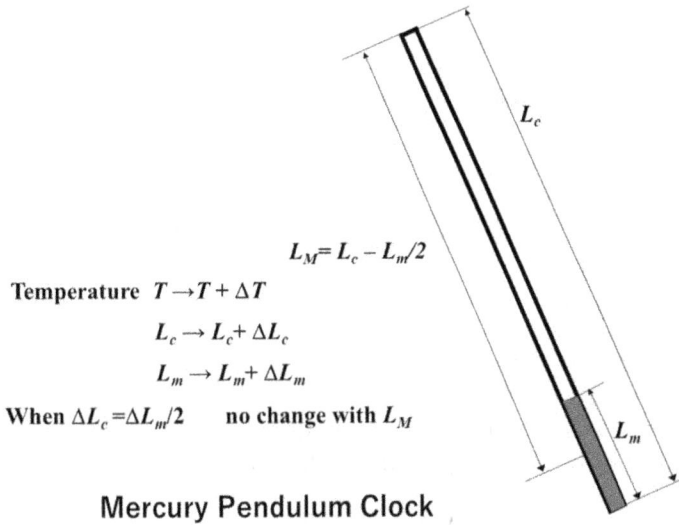

$$L_M = L_c - L_m/2$$

Temperature $T \rightarrow T + \Delta T$

$$L_c \rightarrow L_c + \Delta L_c$$

$$L_m \rightarrow L_m + \Delta L_m$$

When $\Delta L_c = \Delta L_m/2$ no change with L_M

L_c

L_m

Mercury Pendulum Clock

Figure 3.2. Mercury pendulum clock in which the influence of thermal expansion is eliminated.

voltage, and the quartz crystal gives an ac voltage with the vibrational frequency of 1–20 MHz. A quartz clock is based on the oscillation frequency of the quartz crystal, with which the influence of the temperature fluctuation is much less than that with a pendulum clock. For small clocks, a quartz oscillator with a frequency of 2^{15} Hz = 32 768 kHz is often used so that the frequency of 1 Hz is easily counted. The quartz oscillation frequency is much higher than the pendulum oscillation; therefore, the time is measured on a much finer scale. The error is between a few seconds and a few minutes per year. It is currently the most widely used timekeeping technology and has remained so, despite the development of the atomic clock, because of its compact size.

Figure 3.3 shows the history of the standard of time and frequency. The standard of time was based on the solar day, one rotation of the Earth about its axis, and was defined as containing 86 400 s. However, this standard is not always appropriate to describe physical laws, because the solar day is not constant. Over a time-scale of a million years, Earth's rotation is slowing down, because of gravitational tides principally stemming from gravitational interaction with the Moon. It can also fluctuate over time scales of 10–50 years by a factor of 10^{-8} [7].

During the period 1956–67, the standard of time was changed to the period of the Earth's orbit around the Sun. Its fluctuation is on the order of 10^{-9}, which is just one order smaller than the rotation period of the Earth [8]. The orbital period fluctuates due to gravitational perturbations from other planets. Note also that the Sun is losing mass; hence, the radius of the Earth's orbit around the Sun is increasing. Therefore, the orbital period is getting longer.

Although quartz clocks are accurate enough to observe the fluctuation of the solar day, no standard of time and frequency based on quartz clocks has been established. This is because the oscillation frequency of quartz crystal is not a

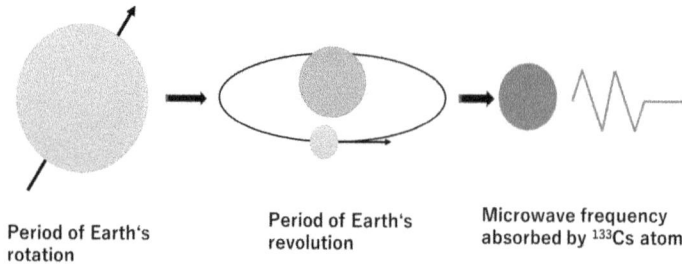

Period of Earth's
rotation

Period of Earth's
revolution

Microwave frequency
absorbed by ^{133}Cs atom

Figure 3.3. History of the standard of time and frequency.

universal value because the properties of different quartz crystals cannot be exactly same.

At 1885, Balmer discovered discrete frequencies f of light emitted from the hydrogen atom following an empirical equation, called the Balmer series [9]

$$f = cR_\infty [\frac{1}{m^2} - \frac{1}{n^2}] \, m \, = 2, \tag{3.2}$$

where m and n are integers, c is the speed of light, and R_∞ is the Rydberg constant for hydrogen atoms. Other series with $m = 1$ (Lyman series) and 3 (Paschen series) were discovered afterwards. As shown in figure 3.4, absorption or emission is possible with discrete frequencies for all kinds of atoms and molecules (although they are not shown with a simple formula like equation (3.2)), and these phenomena were explained with the following quantum mechanical principles.

(1) Each electromagnetic wave (light, microwave, etc) has a characteristic frequency f and an associated particle (called the photon) having energy hf (h: Planck constant).

(2) Each particle has a characteristic wave associated with it. The electrons in atoms can have only orbits, called Bohr orbits, whose length ($2\pi \times$ radius) is a multiple of the electron's matter wavelength (called the de Broglie wavelength). Therefore, the energy of an electron can take only discrete values.

(3) The electrons in atoms can change their state of energy (make a transition) by difference ΔE by absorbing or emitting a photon; for the total energy to be conserved, $\Delta E = hf$. Electromagnetic waves are absorbed or emitted only if the frequency satisfies this condition (called the transition frequency).

After World War II, atomic clocks were invented based on the frequency of the electromagnetic waves absorbed by atoms. At first, atomic clocks were developed based on the atomic transition in the microwave region so that the transition frequency can be measured using a frequency counter. The first clock was developed in 1949 using ammonia molecules, although the improvement in frequency was not so dramatic in comparison with the quartz clock. In 1955, an atomic clock based on the transition frequency of Cesium (Cs) atoms in a thermal beam was constructed at the National Physical Laboratory (UK). At that time, the frequency accuracy was

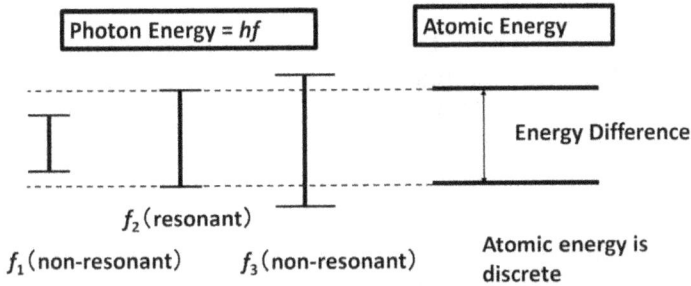

Atoms can absorb or emit photons, whose energy is equal to the energy difference

Figure 3.4. Fundamentals of an atomic clock based on the atomic transition frequency.

10^{-10} (a 1 s error per 300 years). The improvement was achieved for the following reasons.

(i) The atomic transition frequency is discrete in that with a slight shift in the frequency of the electromagnetic wave, the atomic system becomes non-resonant.

(ii) In a gaseous state, atoms are isolated from one another. The characteristics of a single atom are therefore uniform over the gaseous state. In contrast, the bonding state between neighboring atoms in a solid substance cannot be perfectly the same throughout the solid and depends on the conditions (temperature, composition, impurities).

(iii) The atomic energy structure is determined mainly by the Coulomb force between the nucleus and the electrons because, at the atomic scale, the electric field is more than five orders larger than the electric fields that can be generated in the laboratory.

(iv) The atomic transition frequencies are much higher than the oscillation frequency of quartz crystals, and the time is measured with finer scale division.

However, (i)–(iii) are not perfectly satisfied even with atomic clocks. The finite measurement times leads to a broadening of the transition spectrum, which results in statistical uncertainty. The transition frequency is shifted by the electric and magnetic fields, atomic motion, and the gravitational potential, which results in systematic uncertainty (chapter 2).

In 1967, the standard of time and frequency was given by the transition frequency of Cs atoms in the microwave region (9 192 631 770 Hz) measured with a zero electric field, zero magnetic field, zero velocity, and the gravitational potential at the geoid surface. Why was the Cs atomic transition chosen as the standard? The Cs atom is a stable alkali atom with a single valence electron; therefore, the energy structure is relatively simple. Moreover, it has the largest mass (Fr atom is radioactive). With a larger mass, the mean motion velocity becomes slower, and the quadratic Doppler shift given by the relativistic effect is reduced (inversely proportional to mass). Moreover, the transition frequency in the microwave region is the highest for the alkali atoms. Therefore, the atomic transition frequency of Cs was best suited to achieve the highest accuracy [10, 11].

With the present commercial Cs atomic clock, the uncertainty is on the order of 10^{-13} (a 1 s error per 300 000 years), With the laboratory-type atomic clock, with which we can estimate the shifts induced by some causes (measurement velocity distribution in the atomic beam, magnetic field, etc), measurement uncertainty is reduced to the order of 10^{-14}.

Since the development of the laser light source, the measurement uncertainty of time and frequency has been reduced drastically.

3.4 Length standard (until wavelength of light emitted from Kr)

The concept of length was clear in ancient times because it was easy to compare the lengths of different bars. Soon after the French revolution, the length of one meter was defined as one ten-millionth of the distance between the North Pole and the equator [12]. In 1799, a standard bar of platinum was made based on the meridional meter length. This standard bar became known as the 'Mètre de Archives'. In 1869, the definition of the meter was changed to this 'Mètre de Archives' because it was difficult to measure the size of the Earth (estimated from the distance between Dunkirk and Barcelona). It was realized that it was shorter than the defined length by 0.2 mm. The uncertainty was on the order of 10^{-5}. The uncertainty was due to thermal expansion, and the bending of the bar was insignificant.

In 1889, the definition was changed to the distance between two lines on a standard bar measured at the melting point of ice to minimize the effect of thermal expansion. The standard bar was made of an alloy of platinum with 10% iridium (the alloy being harder than pure platinum) with an X cross-section to minimize bending. The uncertainty attributed to thermal expansion, bending, and the thickness (1 μm) of the lines is on the order of 10^{-7}. There is change in the length over time (non-zero for all solid materials), but this effect was expected not to be detectable for 1000 years.

In the 1940s, the use of the wavelength of light emitted by an atom as the length standard was proposed because the property of atoms in the gaseous state is universal, as mentioned in section 3.3. In 1960, a new definition of the meter length was given in terms of the wavelength ($6.057\ 802\ 11 \times 10^{-7}$ m) of light emitted by krypton-86 (^{86}Kr); thus, the uncertainty was reduced to 10^{-9} [13].

The development of the laser was also a significant step in the establishment of a length standard.

3.5 Mass standard

In ancient times, the concept of mass was equivalent to 'weight'. The masses of different objects were compared using a steelyard. Archimedes developed a method to compare the densities of two objects with the same mass by comparing their weights using a steelyard in water (with different volumes and different buoyancies).

According to Newtonian mechanics, the inertial force and the gravitational force are commonly proportional to the mass. Newton distinguished inertial mass from gravitational mass, and the equality of both masses remained a mystery. This problem was solved by the theory of general relativity.

The original kilogram (kg), the base unit of mass, was defined as the mass of water in a cubic container of size 10 cm × 10 cm × 10 cm (i.e. one litre of water). However, it was not a good standard because the density of water is not universal. Water in rivers or lakes is always mixed with various components; therefore, the density of water is not universal. For example, mineral water includes various minerals, such as salt or sulfur compounds. Once there was a proposal to define a kilogram with the 'water in the Donau River'. It was finally defined with distilled water. Water density has dependence on the temperature; therefore, it was defined with the temperature with the highest density (4 °C). Its uncertainty was given by the uncertainties of the container size, purity of the water, and the temperature.

Since 1889, an artifact defines the standard kilogram (International Prototype of the Kilogram or IPK) stored at the International Bureau of Weights and Measures (BIPM) in the outskirts of Paris [14]. It is a right-circular cylinder (diameter = height, figure 3.5) of 39 mm made from an alloy of platinum (90%) and iridium (10%). The hardness of pure platinum is enhanced by the addition of iridium; moreover, its surface area is minimized by using a right-circular shape. In addition, six copies of the IPK were made, but there is a mass difference from that of the IPK of 10–100 μg [15] because of limits in accuracy of size and mixture rates of iridium during fabrication. In addition, mass differences have drifted over time. The masses have increased with dust in the air settling on its surfaces and have also been reduced from the original values when cleaned. The mass uncertainty is on the order of 5×10^{-8}.

While the standards of time and frequency and length are defined with the properties of atoms on a micro scale, the standard of mass has been defined with a

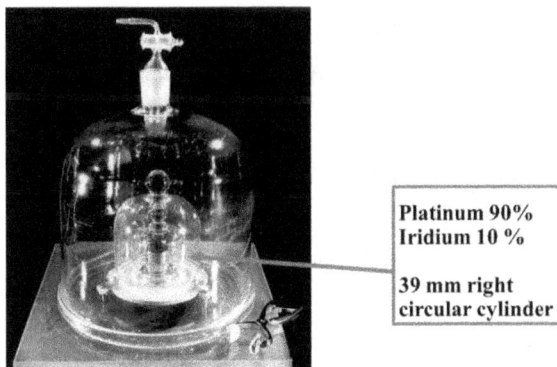

Platinum 90%
Iridium 10 %

39 mm right
circular cylinder

Figure 3.5. The 1 kg mass standard (until May 2019) stored at the International Bureau of Weights and Measures (BIPM). The photo was provided by BIPM, with permission to use it in this book.

material on a macro scale. In November 2018, it was decided to use a new definition of the mass standard beginning in May 2019 (section 5.4).

3.6 Temperature standard

We can feel that a temperature is 'hot' or 'cold', but it has been difficult to measure as a physical value. The first thermometer was made by Galileo at 1592 (Galileo's air thermometer). It consists of a glass bulb about the size of a hen's egg attached to a long, thin tube dipped into a vessel of water, as shown in figure 3.6 [16]. The height of the water surface in the tube becomes lower as the temperature (pressure) in the bulb increases.

Thermometers using the thermal expansion of liquids (mercury, ethanol) were developed afterwards. The scales were evenly spaced based on the observation that thermal expansion is proportional to the change in temperature (discovered by Galileo).

Two units of temperature were proposed. The Celsius temperature scale was defined taking the freezing and boiling points of water as 0 °C and 100 °C. Both of these temperature points are dependent on the pressure, which led to the uncertainty of the definition. In the Fahrenheit temperature scale, 0 °F and 100 °F were defined as the lowest known temperature at that time (−17.8 °C) and normal human body temperature (37.8 °C). The Celsius temperature scale is used in most of world, but in the US, the Fahrenheit temperature scale is used.

With the SI-system, the thermo-dynamical temperature with the unit of Kelvin (K) is used. With this system, 0 K is the lowest temperature, where the velocities of atoms or molecules in a gaseous state are exactly zero, which is not possible from the fundamental of quantum mechanical requirement (uncertainty principle between the position and momentum etc). The ideal gas law given by

$$PV = N_g k_B T, \tag{3.3}$$

where P, V, N_g, k_B, and T are the pressure, volume, number of atoms or molecules in a gaseous state, Boltzmann constant, and thermal-dynamic temperature, respectively. Equation (3.3) shows that the volume becomes zero at 0 K. The interval of the

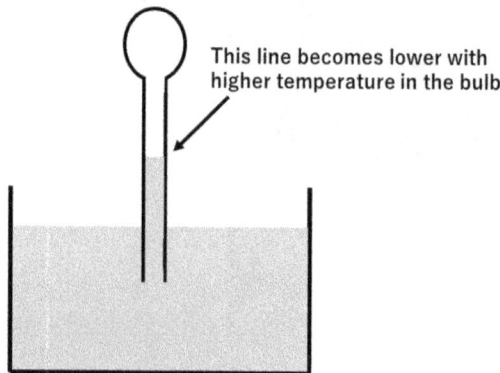

This line becomes lower with higher temperature in the bulb

Figure 3.6. Structure of Galileo's air thermometer.

thermal-dynamic temperature scale is the same as that of the Celsius scale, and it has been empirically shown that 0 K corresponds to −273.1 °C Now the thermal-dynamical temperature is defined taking the triple point of water to be 273.16 K, the uncertainty for which is 0.1 mK. The Boltzmann constant k_B has been measured either using an acoustic gas thermometer or through Doppler broadening of the transition frequency regime of molecules [17]; currently, its uncertainty is less than 10^{-6}. From statistic mechanics, $k_B T$ is the parameter showing the energy distribution. With a thermal equilibrium state, the distribution in each state with energy E is proportional to $\Omega \exp(-E/k_B T)$, where Ω is the degeneracy of states. When Ω is constant, the mean energy is $k_B T$. The mean kinetic energy in one direction is $k_B T/2$, from which equation (3.3) is derived.

3.7 Electric current standard

Electromagnetic theory was very rudimentary until the 17th century, and the phenomenon of lightening in a thunderstorm was a mystery. In 1785, Coulomb's law was discovered, which was the first law of electromagnetism. In 1800, the induction of a magnetic field by an electronic current was discovered, and Ampere's circuital law was discovered in 1823. These laws were finally summarized by Maxwell's equation [18] between 1861 and 1862.

Electronic current is a flow of electric charge, and the unit of Ampere (A) is used to mean the flow rate of the electric charge of one Coulomb per second. However, a measurement of the number of flowing charged particles (electrons) per unit time is not realistic. Therefore, electric current has been commonly measured using the induced magnetic field. The electric current of 1 A has been defined as the 'constant current that produces a force of attraction of 2×10^{-7} Newtons (kg m s^{-2}) per meter of length between two straight parallel conductors of infinite length and a negligible cross-section with the separation of 1 m (figure 3.7)'. In reality, a conductor must have a finite length and a non-zero cross-section. Therefore, the ultralow uncertainty of an electric current with this definition is not possible, but it satisfied the requirements for physics and engineering research until the 20th century. Recently, microchip electric circuits with small electric current were developed, and the electric current can be determined from the number of flowing charged particles (for example, electrons). Then a definition of electric current with lower

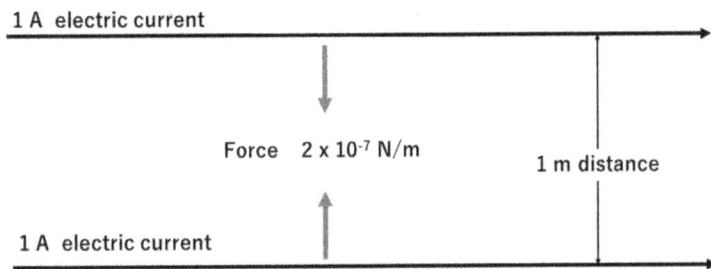

Figure 3.7. Definition of electric current of 1 A using the force induced by a magnetic field.

uncertainty was required, and a new definition was introduced in November 2018 (section 5.6).

3.8 Luminous intensity standard

Luminous intensity is a parameter to express the brightness of a light source, in terms of human vision. The power of irradiated light is given in units of W ($= \mathrm{J\ s^{-1}} =$ kg m^2 s^{-3}). However, human eyes can see only visible light. Moreover, sensitivity to light with a given power depends on the color (highest sensitivity with green-yellow light at 555 nm). Luminous intensity is a measure of the frequency-weighted power in a given direction per unit solid angle (figure 3.8) [19]. The unit of luminous intensity is the candela (cd), and one candela is the brightness of a typical candle. For the SI-unit, one candela is defined as the 'brightness of light with a frequency of 540 THz (wavelength 555 nm) with a radiant intensity if 1/683 W per steradian'. The luminous intensity is increased in one direction by focusing the light. Changing the light frequencies with a constant radiant intensity, the luminous intensity becomes lower, and it becomes zero in the infrared or ultraviolet region.

3.9 Substance quantity standard

Substance quantity is the parameter to give the number of atoms or molecules on a macro scale. It is expressed in units of mol from (number of atoms and molecules/ N_A), where N_A is the Avogadro constant. Here, N_A is defined as the 'number of ^{12}C atoms contained in a 12 g (0.012 kg) of carbon'. The recommended value of N_A in 2014 was 6.022 140 857(74) × 10^{23} [20]. The mass of atoms or molecules is roughly obtained by (atomic and molecular weight/1000 N_A) with the unit of kg. Recently, N_A has been measured from the number of atoms in a crystal from the combination of measurement of the lattice constant by x-ray diffraction and size measurement by laser interferometry. The uncertainty of this measurement comes from the mixture of other atoms in the crystal, lattice defects, and uncertainty at the mass

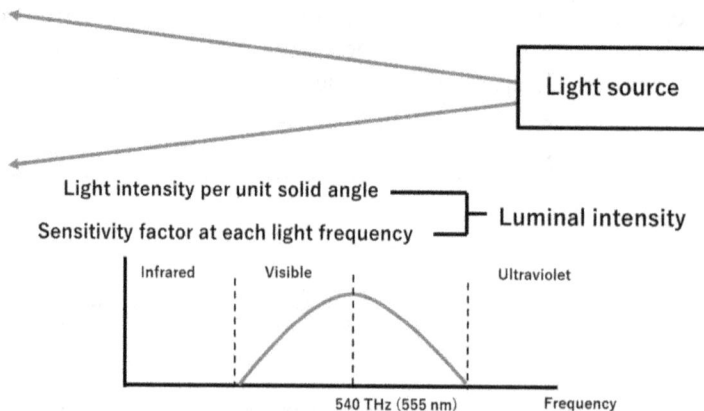

Figure 3.8. Concept of luminal intensity.

measurement of the crystal. A new definition of N_A was introduced in November 2018 (section 5.5).

References

[1] https://sciencemadesimple.com/metric_system.html
[2] Moss T 2013 *How do Sundials Work* (London, UK: British Sundial Soc.) https://web.archive.org/web/20130802033857/ http://sundialsoc.org.uk/HDSW.php
[3] Goodenow J, Orr R and Ross D 2007 *Mathematical Model of Water clocks* (Rochester, NY: Rochester Institute of Technology)
[4] Usher A P 1988 *A History of Mechanical Inventions* (New York: Dover) p 194
[5] Milham W I 1945 *Time and Timekeeping* (New York: MacMillan) pp 330 334
[6] Milham W I 1945 *Time and Timekeeping* (New York: MacMillan) p 193
[7] *The Astronomical Almanac Online* http://asa.usno.navy.mil/SecM/Glossary.html#_L
[8] *Orbital period* https://en.wikipedia.org/wiki/Orbital_period
[9] Nave C R 2006 Hydrogen spectrum *HyperPhysics* (Atlanta, GA: Georgia State University)
[10] Vanier J and Audoin C 1989 *The Quantum Physics Atomic Frequency Standards* (Bristol: Adam Hilger) p 610
[11] Vanier J and Tomescu C 2017 *The Quantum Physics Atomic Frequency Standards: Recent Developments* (Boca Raton, FL: CRC Press) p 1
[12] https://en.wikipedia.org/wiki/Metre
[13] Baird K M and Howlett L E 1963 *Appl. Opt.* **2** 455
[14] *Resolution of the 1st CGPM* 1889 https://bipm.org/en/CGPM/db/1/1/
[15] Jabbour Z J and Yaniv S L 2001 *J. Res. Natl. Inst. Stand. Technol.* **106** 26
[16] http://scienceuniverse101.blogspot.com/2012/02/galileos-air-themometer.html#.W72lAEkUlaQ
[17] Fischer J 2016 *Philos. Trans. R. Soc.* A **374** 20150038
[18] https://en.wikipedia.org/wiki/Maxwell%27s_equations
[19] https://en.wikipedia.org/wiki/Luminous_intensity
[20] CODATA value: Avogadro constant. NIST 2015

Chapter 4

Lasers revolutionized physics

4.1 Introduction

The invention of lasers made it possible to reduce the measurement uncertainties of transition frequencies of atoms and molecules, which led to a significant revolution in physics (chapters 5 and 6). This chapter introduces the special properties of lasers as light sources with a uniform phase and narrow frequency linewidth. The significant roles of lasers in the development of new research fields are also presented. These include observation of the spectra of atoms and molecules, measurement of light frequencies, and laser cooling of atoms and molecules.

4.2 Fundamentals of lasers

The word 'LASER' is an acronym of '**L**ight **A**mplification by **S**timulated **E**mission of **R**adiation'. This has been demonstrated by quantum mechanics.

(1) All particles have the characteristics of waves. On the other hand, electromagnetic waves have the characteristics of particles. Energy is given by $E = hf$ (h: Planck constant, f: frequency) and their momentum by $p = h/\lambda$ (λ: wavelength). As there is no significant change in the amplitude of the wavefunction within the phase difference of ± 1 (phase has no physical appearance), there is an uncertainty principle between 'time and energy' and 'position and momentum'. When a particle is localized in an area of size L, the wavelength can only assume discrete values of $\lambda = 2L/$(integer); hence, this is also true for the energy and momentum of the particles.

(2) The energy of electrons in an atomic state can be changed ($E_1 \rightarrow E_2$) by the absorption or emission of a photon with an energy equal to the energy difference between the initial and final states (atomic transition energy). The total energy before and after the transition must be conserved. Therefore, atoms can only absorb or emit electromagnetic waves of a specific frequency satisfying $f = |E_1 - E_2|/h$, known as the transition frequency or resonance

doi:10.1088/2053-2563/ab0373ch4

frequency. The total momentum of atoms and photons before and after the transition must be also conserved in all directions.

Transition occurs between two energy states 1 and 2, having energies of E_1 and E_2 ($E_1 < E_2$) and the populations $\rho(1)$ and $\rho(2)$. For simplicity, the degeneracies at both states are assumed to be 1. By irradiation of light with the transition frequency, transition between the two states is induced with the ratio of $BI(\rho(1) - \rho(2))$, where B is the coefficient, and I is the intensity of the incident light wave. With a thermal equilibrium state, $\rho(1) > \rho(2)$ and the incident light wave is damped, as shown in figure 4.1. There is also spontaneous emission transition at the rate of $A\rho(2)$. With spontaneous emission, light is emitted as fluorescence having random phase and direction. The equilibrium population is given by $BI(\rho(1) - \rho(2)) = A\rho(2)$. When I is given by blackbody radiation, $\rho(2)/\rho(1) = \exp[-(E_2 - E_1)/k_B T]$.

When $\rho(1) < \rho(2)$ (called inversion population), the incident wave is amplified with the same phase and direction as those of the incident light wave. Therefore, a substance having an inversion population is used as the light amplifier, which is the fundamental of the laser. Pumping to energy state 2 (some artificial treatment to increase $\rho(2)$) is required to obtain the inversion population between states 1 and 2. It is realistic to consider a three-state model with states 0, 1, and 2 with the energies of E_0, E_1, and E_2 ($E_0 < E_1 < E_2$) and populations of $\rho(0)$, $\rho(1)$, and $\rho(2)$. When the population in state 0 is much higher than that in state 1, the inversion population between states 1 and 2 is obtained by $0 \rightarrow 2$ pumping with a realistic pumping rate, as shown in figure 4.2.

A laser medium can amplify any incident light waves (spontaneous emission, blackbody radiation, etc), but their effects are generally negligibly small except when the laser medium is contained in a cavity. When the amplification (gain) of the light given by the laser medium is higher than the loss inside the cavity, the light with the resonant mode to the cavity repeats amplification while reflecting in the cavity. As the light intensity is increased, $\rho(2) - \rho(1)$ becomes smaller due to the higher stimulated emission rate (saturation effect). Then the gain becomes smaller and

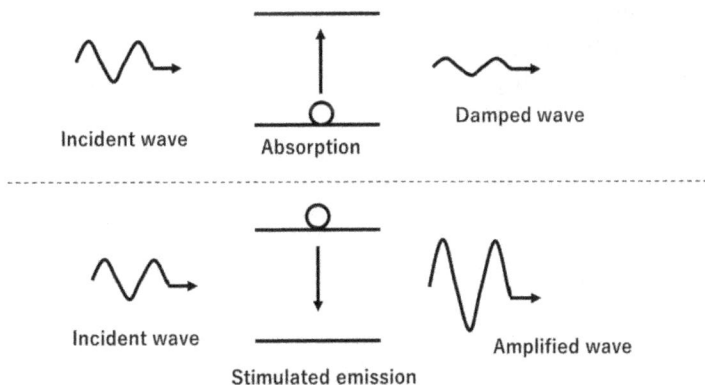

Figure 4.1. Damping and amplification of the incident light wave induced by absorption or stimulated emission by atoms or molecules.

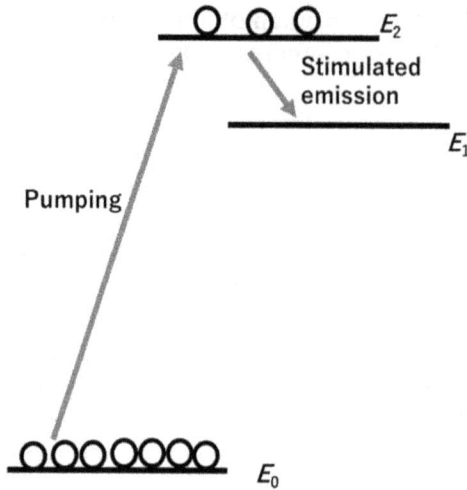

Figure 4.2. Inversion population between states 1 and 2 by $0 \rightarrow 2$ pumping.

amplification stops with the light intensity where the gain balances with the cavity loss. Through this mechanism, a cavity containing a laser medium works as a laser oscillator. Figure 4.3 shows an example using a Fabry–Perot cavity.

The wavelength of the light resonant to the Fabry–Perot cavity with the length of L is $2L/n_m$, where n_m is the integer; therefore, the resonance frequency is $n_m c/2L$ (c: speed of light). With the detuning from the resonance frequency of Δf, there is a phase shift of $\delta = 2\pi L \Delta f/c$ with one reflection. The resonance spectrum of the Fabry–Perot cavity is obtained by $\sum_{k=0}^{N_r} e^{ik\delta}$ (N_r: mean reflection time). The resonance linewidth is obtained to be $c/2N_r L$, which converges to zero with N, $L \rightarrow \infty$, as shown in figure 4.4.

Considering that each photon remains in the cavity while it reflects N_r times (duration of $2N_r L/c$), the spectrum linewidth is the loss rate of the light in the cavity. The Q-value is defined by (frequency/loss rate) $= n_m N_r$, which corresponds to the number of light oscillations while it remains in the cavity.

When laser oscillation is possible only with the single cavity resonance frequency, the laser oscillates with a single frequency with a narrow linewidth. To select a single resonance frequency, a grating mirror (high reflectivity only with a limited frequency area) is often used.

The special properties of light obtained by laser oscillation are the following.
 (i) Uniform phase; therefore, interference is easily observed.
 (ii) Propagation in a single direction parallel to the cavity.
 (iii) Narrow spectrum linewidth at each frequency component.
 (iv) Ultra-short pulse obtained from the interference between many frequency components.
 (v) Focusing to the wavelength size is possible.

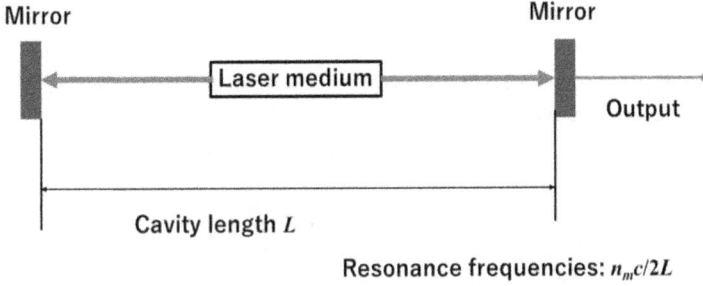

Figure 4.3. Laser oscillator using a Fabry–Perot cavity containing a laser medium. Here, c is the speed of light, and n_m is the integer.

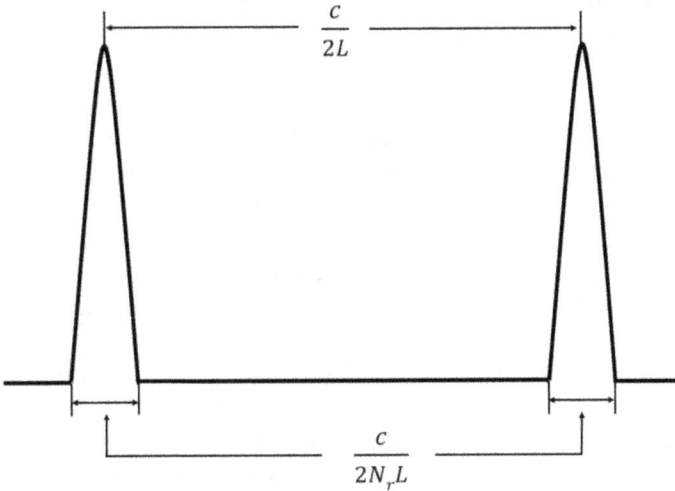

Figure 4.4. Resonance spectrum of a Fabry–Perot cavity with the length of L and mean reflection time of N_r, where c is the speed of light.

4.3 Laser spectroscopy

Measurement of the transition frequencies of atoms or molecules is useful to investigate their quantum energy structures. A high transition rate is obtained using a light source with a narrow frequency bandwidth, and laser light is useful for this purpose. The uniform propagation direction of laser light is also advantageous because it makes control of the optical path easy.

Since 1970, the search for polar molecules has been performed using CO_2 or N_2O lasers, with which there are many discrete transition frequencies. Frequency scanning is not possible with these lasers; therefore, the spectra were observed by inducing the transition frequencies to shift by the application of a DC electric field (laser Stark spectroscopy). The spectrum was first observed for NH_3 molecules, having a simple energy structure and a high absorption coefficient [1]. This method has also been applied to other molecules (H_2CO [2], CH_3F [3] etc). Later, frequency-tunable dye lasers and diode lasers were developed, and the observation of spectra by frequency scanning became possible [4]. The frequency change by scanning was monitored using a Fabry–Perot interferometer.

Until the frequency comb was developed (section 4.4), frequency measurement was performed using wavelength meters, with which the attained accuracy was on the order of 10^{-7}. However, the frequency difference between two laser lights was measured by observing the beat signal.

The observed spectrum has finite broadenings: lifetime broadening and Doppler broadening. Lifetime broadening is the broadening that occurs due to the limited time of interaction between light and atoms or molecules without a phase jump, which is attributed to the uncertainty principle between time and energy. With the mean interaction time of τ_e, the broadening of $1/2\pi\tau_e$ is induced. Doppler broadening is induced by the Doppler shift, which is attributed to the motion of atoms or molecules in the direction parallel to the laser light.

Saturation spectroscopy is useful to suppress Doppler broadening [5]. Pump (high intensity) and probe (low intensity) lasers with the same frequency and counter-propagating directions are shown in figure 4.5. Saturation reduces the absorption coefficient of the probe laser induced by the pump laser, which decreases the population difference between the upper and lower states. When the probe laser is resonant with the atoms or molecules with the velocity of v, the pump laser is resonant with atoms or molecules with the velocity of $-v$. With high velocity, the effect of the saturation induced by the pump laser is not observed. However, the saturation effect is significant with $v = 0$, and a dip is observed on the absorption line (see figure 4.5).

The transition induced by the laser light can be monitored from the damping of laser light, as shown in figure 4.5. This method is applicable when the number of atoms and molecules is large enough that the damping is greater than the power fluctuation of the laser light. The detection of fluorescence induced by transition is useful for transitions in which the spontaneous emission transition rate is high. The fluorescence detection sensitivity is much higher than that for the damping.

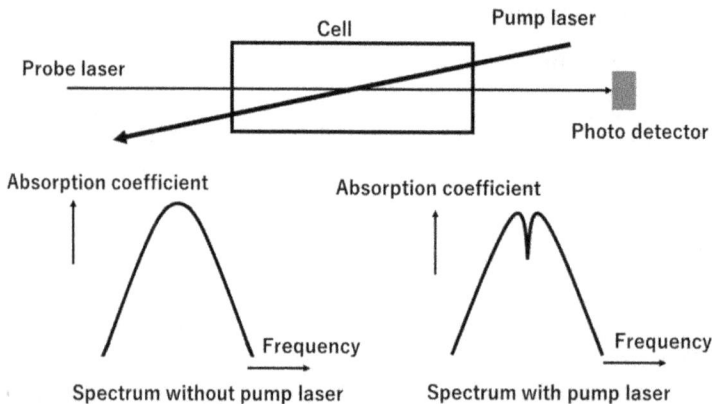

Figure 4.5. Measurement of the saturation spectrum using pump and probe lasers. With the saturation effect caused by the pump laser, a dip is observed at the center of the Doppler broadened spectrum.

However, transitions with high spontaneous emission rates are not advantageous for precise frequency measurement because of the large lifetime broadening (phase of atomic or molecular wavefunction jumps with short time by spontaneous emission). Double resonance is often used for the precise measurement of atoms or molecules with small numbers. A detection laser is also irradiated so that the atoms or molecules in one quantum state can be detected by the fluorescence (see figure 4.6). Then we can monitor the transition induced by the probe laser. The transition spectra of atoms and molecules are also used for the stabilization of the laser frequency.

4.4 Measurement of laser frequency

Transition frequency measurement with an uncertainty lower than that obtained with the wavelength meter (order of 10^{-7}) is required to get detailed information regarding the energy structures of atoms and molecules. The fundamental of frequency measurement is measurement of the frequency difference from the beat signal.

To measure the optical frequency, the frequency chain system was developed (figure 4.7) [6]. The fundamental of the frequency chain is the higher harmonic oscillation of the crystals. When an ac electric field $E = E_0 \cos(2\pi ft)$ is applied to a crystal, oscillation of the charge distribution is induced. The waveform of this oscillation is distorted in comparison to that of the incident electric field, and the frequency components of $2f$, $3f$, $4f$,..., are included. Therefore, the output radiation from the crystal includes the frequency components of integer multiples of the frequency of the incident wave. In the frequency chain, an integer multiple of the frequency of the lower-frequency source is used to lock the frequency of the higher-frequency source. The standard frequency given by the Cs atomic clock is multiplied and locked to the far-infrared laser (FIR) by measuring the beat signal frequency. Then the FIR frequency is also integer multiplied and locked to the infrared laser

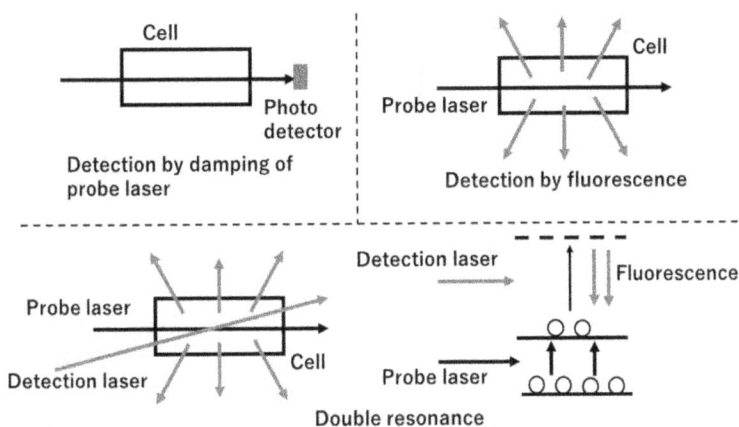

Figure 4.6. Method to monitor the transition of atoms or molecules induced by the probe laser by damping of probe laser or by observation of fluorescence. Double resonance is often used to also irradiate a detection laser to detect the fluorescence from atoms or molecules in a selected state.

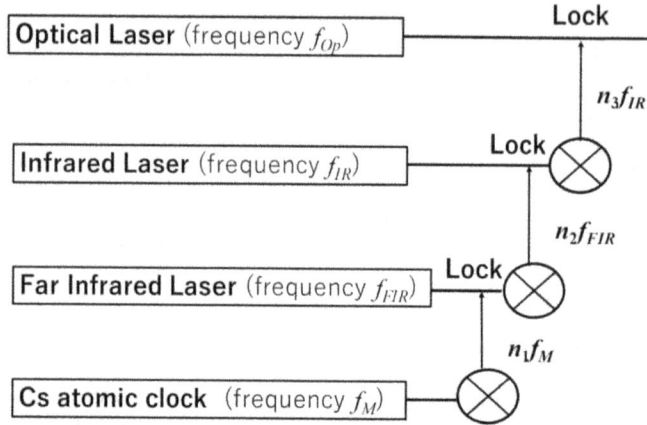

Figure 4.7. Schematic of the structure of a frequency chain constructed using a Cs atomic clock, a far-infrared laser, an infrared laser, and an optical laser.

(IR). With multiplication of the IR frequency, the laser light frequency in the optical region can be locked, and its frequency can be measured. The operation of the frequency chain is rather complicated because several frequency sources must operate simultaneously.

At the beginning of the 21st century, the frequency comb system was developed [7]. As mentioned in section 4.1, the resonance frequency of a cavity with a length of L is $n_m c/2L$, where n is an integer. When laser oscillation is possible for the frequency components of $N_0 - N_c \leqslant n_m \leqslant N_0 + N_c$ with a uniform phase, the laser light is a pulse laser with the length of $2L/N_c c$, which repeats with the frequency of $f_{rep} = c/2L$. The cavity loss is controlled so that laser oscillation is possible only with a uniform phase for all frequency components (mode locking). The reciprocal of the repetition rate corresponds to the period of a round trip in the cavity of the light. The frequency components are sum of integer multiples of the repetition frequency f_{rep} and an offset frequency f_{ceo}, which can be easily measured (figure 4.8). The laser light frequency can be measured by interferometry using the closest frequency component $f(n_m) = f_{ceo} + n_m f_{rep}$. In comparison with a frequency chain, a frequency comb is much more compact and much simpler in operation. The frequency components can be stabilized by locking f_{rep} and f_{ceo} using an atomic clock in the microwave region. However, it is more common to lock one frequency component using an optical atomic clock, and then all frequency components are locked simultaneously. As will be seen in chapter 5, frequency measurement with uncertainty on the order of 10^{-18} is possible using a frequency comb.

The output of a frequency comb has a short pulse, which is possible because of the presence of many frequency components. However, the linewidth of each frequency component is narrow; therefore, it can be used not only for frequency measurement of the probe laser, but also as the probe laser itself. For example, a frequency comb is useful to observe the Raman transition (two-photon transition with the frequency of f_R induced by two laser lights with frequencies of f_a and $f_a - f_R$). Raman transition

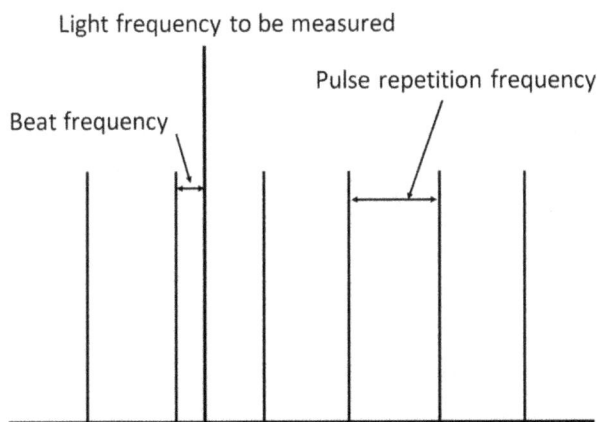

Figure 4.8. Schematic of the frequency spectrum of the frequency comb.

is induced when $f_R = N_R f_{rep}$ (N_R: integer) induced by frequency components of $n = n_c$ and $n_c - N_R$ (figure 4.9). Many frequency components contribute to inducing the Raman transition because there are many possible values of n_c. A frequency comb is also useful for two-photon absorption. When the transition frequency f_t satisfies $f_t = 2f(n_t)$, this transition is induced also by the combination of the frequency components of $f(n_t + n')$ and $f(n_t - n')$. Therefore, many frequency components also contribute to two-photon absorption (figure 4.9). Two-photon absorption is useful to observe spectra without Doppler broadening.

A dual-frequency comb is useful to observe many transitions (particularly molecular transitions) simultaneously, as shown in figure 4.10. We consider frequency combs A and B with the following frequency components:

$$\text{Comb A} f_A(n_c) = f_{ceo} + n_c f_{rep},$$
$$\text{Comb B} f_B(n_c) = f_{ceo} + \delta f_{ceo} + n_c(f_{rep} + \delta f_{rep}).$$

When the n_c frequency component of comb A is absorbed by atoms or molecules, its power loss is observed as a heterodyne signal with the n_c frequency component of comb-B of frequency $\delta f_{ceo} + n_c \, \delta f_{rep}$. From the frequency of the heterodyne signal, we know which frequency component of the comb was absorbed. When the n_{c1}th and n_{c2}th frequency components of comb A are simultaneously absorbed, the heterodyne signal is constructed by the frequency components of $\delta f_{ceo} + n_{c1}\delta f_{rep}$ and $\delta f_{ceo} + n_{c2} \, \delta f_{rep}$, which can be resolved by mass spectrometry. With this method, many transitions can be observed with a single measurement [8].

4.5 Laser cooling

Another revolution was brought about in atomic physics by lasers. The development of laser cooling technology made it possible to decelerate gaseous atoms from the speed range of 200–400 m s^{-1} to slower than 10 cm s^{-1}. With ultra-low kinetic energy (order of 10^{-6} K), the properties of atoms as waves (for example atomic interferometry [9]) become much more significant than at room temperature. As will

$n_1 f_{rep} + f_{ceo}$ $(n_1 - N_R) f_{rep} + f_{ceo}$

$n_2 f_{rep} + f_{ceo}$

$(n_2 - N_R) f_{rep} + f_{ceo}$

$N_R f_{rep}$

$n_I f_{rep} + f_{ceo}$ $(n_I - n') f_{rep} + f_{ceo}$

$n_I f_{rep} + f_{ceo}$ $(n_I + n') f_{rep} + f_{ceo}$

Raman transition   **Two-photon absorption**

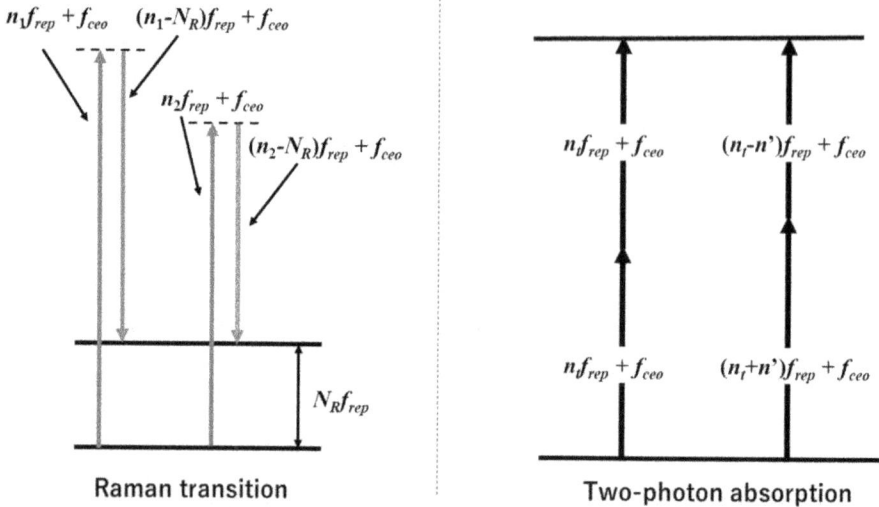

Figure 4.9. Raman transition and two-photon absorption using many frequency components of a frequency comb.

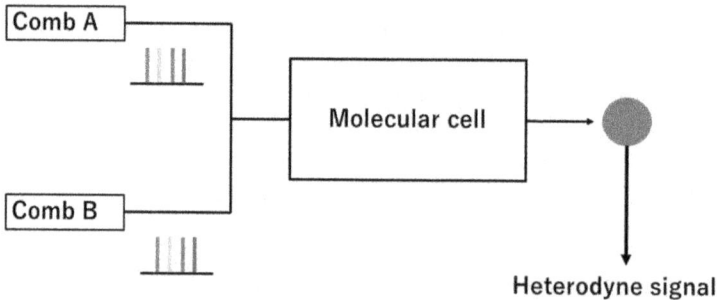

Comb A

Molecular cell

Comb B

Heterodyne signal

Figure 4.10. Schematic of measurement using dual-frequency combs.

be discussed in section 5.2, laser cooling contributed greatly to the reduction of measurement uncertainty of atomic transition frequencies, reducing the lifetime broadening and the quadratic Doppler shift.

Here, we consider an atom with one ground state and one excited state. When atoms absorb laser light, they absorb both the energy and the momentum of the photons. Therefore, atoms are transformed to an excited energy state and simultaneously experience a force along the direction of light propagation [10]. Then the atoms emit photons in random directions (spontaneous emission) and return to the ground state. With the emission of photons, atoms receive a recoil force; nevertheless, it is randomly directed and averages to zero. Repeating the cycle of absorption and spontaneous emission, atoms experience a net force along the direction of light propagation. To cool atoms of random velocities, two laser beams with frequencies lower than the transition frequency irradiate the atoms from opposing directions. Because of the first-order Doppler effect, atoms interact only with opposing light; all atoms are decelerated (figure 4.11). As the atoms decelerate and the Doppler shift becomes smaller than the lifetime broadening (Γ), the difference between the interaction forces of the two opposing beams becomes smaller. Cooling stops when the kinetic temperature, called the Doppler limit, is reached; its minimum value is given by $T_{\text{Doppler}} = h\Gamma/2k_B$, where h is the Planck constant, and k_B is the Boltzmann constant.

We assumed that only spontaneous emission transition from the excited state to the initial ground state is possible. When a second transition to another state is possible, the cooling cycle (absorption + spontaneous emission) cannot continue for long. A repump laser, causing transition to the excited state (figure 4.12), is needed to push atoms back to the cooling cycle. With complex energy states, more repump lasers are required to maintain the laser cooling process. Therefore, laser cooling is possible only for atoms with simple energy structures, such as alkali or alkali-earth atoms. Recently, laser cooling was successfully applied to several molecules with relatively simple energy structures [11, 12]. In a variant of this technique, called sympathetic cooling, the kinetic energy of atoms or molecules can be reduced via collisional interactions with laser-cooled atoms. Sympathetic cooling is useful to reduce the kinetic energy of atoms or molecules for which direct laser cooling is difficult.

When laser cooling is performed in one direction, there is a heating effect in the other direction induced by the random force during spontaneous emission. To reduce the kinetic energy in two or three directions, four or six cooling lasers are required.

For ions trapped by an electric field, laser cooling can be performed with one cooling laser that has a frequency lower than the transition frequency. Trapped ions repeatedly undergo motion parallel and opposite to the direction of light and

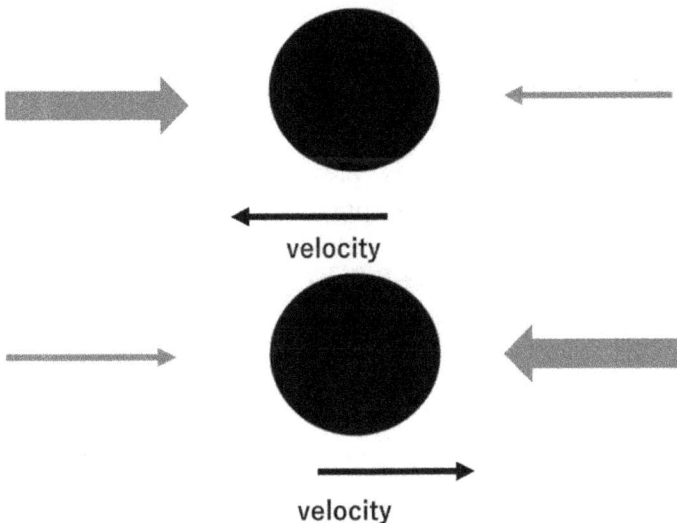

Figure 4.11. Laser beams with frequencies below the transition frequency irradiate the atoms from opposing directions. Atoms interact only with laser light that opposes their motion.

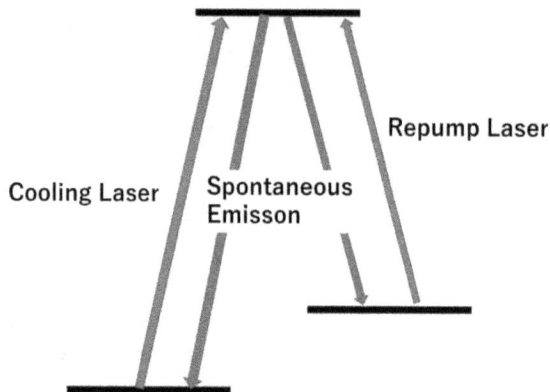

Figure 4.12. Maintaining the cooling cycle using a repump laser.

interact with the cooling laser only if the motion is in the opposite direction to the beam. The motions of the trapped ions in the three directions are coupled and cooled in all directions by a cooling laser in one direction.

Kinetic energy lower than the Doppler limit can be obtained by several methods, such as polarization gradient cooling [13], sideband cooling [14], sideband Raman cooling [15] and so forth.

References

[1]   Shimizu F 1970 *J. Chem. Phys.* **52** 3572
[2]   Johns J W C and McKellar A R W 1973 *J. Mol. Spectrosc.* **48** 354
[3]   Freund S M *et al* 1974 *J. Mol. Spectrosc.* **52** 38
[4]   Job V A *et al* 1983 *J. Mol. Spectrosc.* **101** 48
[5]   Preston D W 1996 *Am. J. Phys.* **64** 1432
[6]   Evenson K M *et al* 1973 *Appl. Phys. Lett.* **22** 192
[7]   Adler F *et al* 2004 *Opt. Express* **12** 5872
[8]   Shimizu Y *et al* 2018 *Appl. Phys.* B **124** 71
[9]   Baudon J *et al* 1999 *J. Phys. B: At. Mol. Opt. Phys.* **32** 201
[10] Haensch T W and Shawlow A L 1975 *Opt. Commun.* **13** 68
[11] Griffith W C *et al* 2009 *Phys. Rev. Lett.* **102** 101601
[12] Kozyryef I *et al* 2017 *Phys. Rev. Lett.* **119** 133002
[13] Dalibard J and Cohen-Tannoudji C 1989 *J. Opt. Soc. Am.* B **6** 2023
[14] Morigi G *et al* 1999 *Phys. Rev.* A **59** 3797
[15] Hamann S E *et al* 1998 *Phys. Rev. Lett.* **80** 4149

IOP Publishing

Measurement, Uncertainty and Lasers

Masatoshi Kajita

Chapter 5

Revolution of measurement uncertainties due to the introduction of lasers

5.1 Introduction

This chapter discusses the significant reduction of measurement uncertainties after the invention of lasers (section 4.2). Using laser light with a narrow frequency linewidth, we can observe atomic transition with high sensitivity (section 4.3); then we can observe atomic transition with a low absorption coefficient and low spontaneous emission transition rate, which are advantageous for precise measurement because of the narrow spectrum linewidth (shown in equation (2.8)). The frequency of a probe laser can be measured using a frequency comb (section 4.4). The measurement uncertainty is drastically reduced using laser-cooled atoms (section 4.5).

Through the precise measurement of atomic transition frequencies, the uncertainties of fundamental physical constants were reduced. After the definition of several fundamental constants without uncertainty, the uncertainties of all physical values are reduced.

5.2 Measurement of time and frequency using lasers

5.2.1 Reduction of measurement uncertainty of Cs standard

Currently the standard of time and frequency is determined by the transition frequency in the microwave region of Cs atoms with zero electric field, zero magnetic field, zero velocity, and on the geoid plane. It is not realistic to satisfy these conditions, and the Cs atomic clocks in the laboratory were operated with correction of the estimated frequency shifts induced in various ways. The uncertainty of the frequency correction is systematic uncertainty. There is also a statistical uncertainty due to the limited interaction time broadening between the Cs atoms and microwaves. The lowest measurement uncertainty using an atomic clock using atomic beams at room temperature is 7×10^{-15} [1]. It is difficult to achieve lower uncertainty because of the high velocity of Cs atoms (of the order of 300 m s^{-1}). The frequency shift due to the relativistic effect (quadratic Doppler shift) at this velocity is -5×10^{-13}. The limited interaction time between Cs atoms and microwaves results in a spectrum linewidth on the order of 100 Hz using a microwave cavity with the length of 1.5 m (with commercial type, a few kHz).

The development of laser cooling technology (section 4.5) allowed the velocity of Cs atoms to be lowered to a few centimetres per second. Using cold atoms, the interaction time between atoms and microwaves is longer, and the linewidth becomes narrower (< 1 Hz). The quadratic Doppler shift is also reduced ($< 10^{-16}$). In 1991, the first fountain-type Cs atomic clock (figure 5.1) was developed at the Systemes de Reference Temps Espace (SYRTE), France [2]. Atoms that have been laser cooled and trapped are launched upward using laser light. The launched atoms pass through a microwave cavity, reverse direction, and fall through the cavity again. Cs atomic clocks using an atomic fountain were also constructed in the US, Germany, the UK, Italy, Japan, and China. A fractional uncertainty on the order of 10^{-16} was obtained by several groups [3]. A ^{87}Rb atomic clock with an atomic fountain was also developed with which a fractional uncertainty on the order of 10^{-16} was obtained [4].

In the 1990s, the continuous operation of an atomic clock with an atomic fountain was difficult. The primary atomic clocks in laboratories (beam or atomic fountain) were utilized to evaluate the accuracy of the national standard time produced by commercial atomic clocks. Nowadays, atomic clocks with atomic fountains can be operated continuously. Physikalische Technische Bundesanstalt (PTB, Germany), the National Institute of Standard Technology (NIST, US), and Systemes de Reference Temps Espace (SYRTE, France) provide national standard times mainly using Cs atomic clocks with atomic fountains.

The development of chip-scale diode lasers made it possible to make a chip-scale atomic clock (CSAC). The microwave transition frequency between a and b states

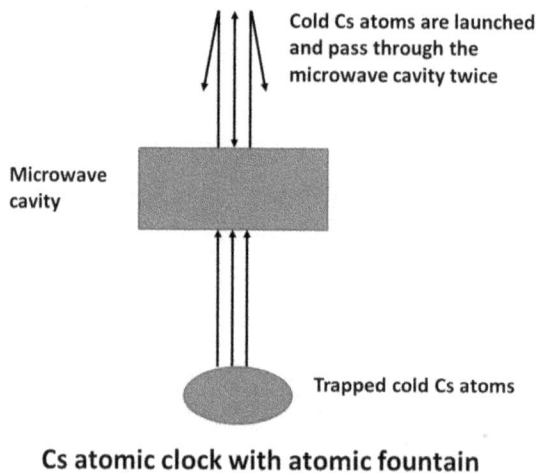

Cs atomic clock with atomic fountain

Figure 5.1. Structure of a Cs atomic clock with an atomic fountain.

can also be measured using a laser with two frequency components. Laser light is absorbed by atoms when its frequency corresponds to the transition frequency of either a–e or b–e (e is the highly excited state). However, light is not absorbed when there are two frequency components and their frequency interval is equal to the a–b transition frequency; this phenomenon is called electromagnetically induced transparency or EIT (figure 5.2).

Multiple frequency components of a laser are made by modulating the laser frequency; their frequency intervals are given by the modulation frequency. The microwave transition frequency is measured by monitoring the modulation frequency at which the laser light is not absorbed. With the development of the compact laser diode having low power consumption, an atomic clock using an EIT much smaller than the microwave wavelength was developed; this is not possible using microwave technology [5]. Low-cost CSACs with low power consumption can be inserted into many devices, such as mobile phones, personal computers, and GPS receivers.

5.2.2 Toward a new standard of time and frequency using optical transition frequencies

The attainable accuracy of atomic clocks based on optical transition is expected to be higher than that for microwave transitions because the time unit of 1 s can be divided into finer scales over five orders (figure 5.3). However, optical frequencies cannot be measured using a frequency counter; therefore, the microwave transition frequency has been used for the standard of time and frequency. Since the development of the frequency comb (see section 4.4), the measurement of the optical transition frequencies became possible, and the topic of precise measurement of the atomic transition frequencies generated intense interest from researchers.

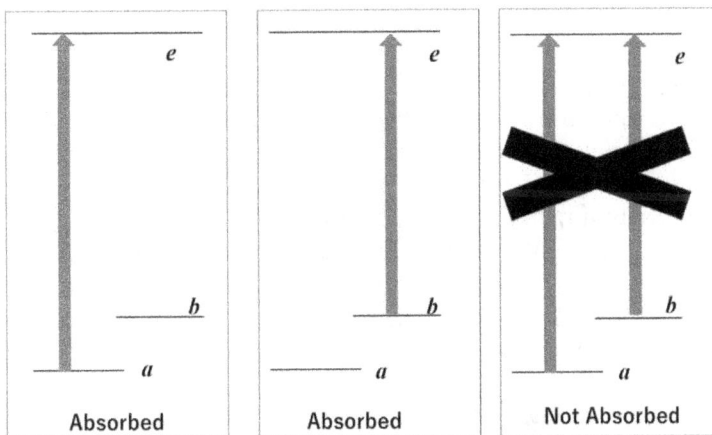

Figure 5.2. Fundamentals of electromagnetically induced transparency (EIT). Irradiating laser light that is tuned to the a–e or b–e transition is absorbed. However, the light from two laser beams tuned separately to the transitions are not absorbed.

Optical measurement

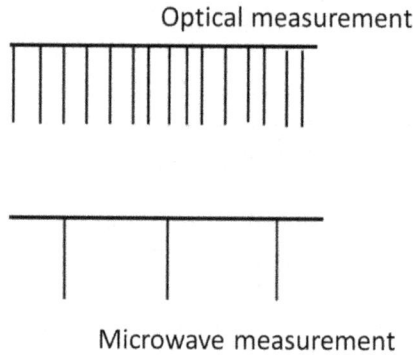

Microwave measurement

Figure 5.3. Comparison of time scales for optical and microwave measurements.

The Garching group measured the 1s–2s transition frequency of the hydrogen (H) atom. The measurement was performed using a H atomic beam emitted from a nozzle cooled by liquid He, so that the velocity of the atomic beam would be reduced (figure 5.4). The transition was observed by two-photon absorption induced by two laser beams (wavelength 243 nm) incident from opposing directions so that the Doppler effect would be eliminated. The angle between the laser light and the atomic beam was set as small as possible to ensure a long atom–light interaction time (narrow linewidth). In 2000, this uncertainty in measurement was reduced to 1.8×10^{-14} using a Cs clock with an atomic fountain as a reference [6]. To date, lower measurement uncertainties have been obtained with some other transition frequencies (see below). However, H atoms are the only atoms for which the energy structure is calculable and precise measurements of the H atomic transition frequencies are useful in advancing fundamental physics. For example, the uncertainty of the Rydberg constant was reduced through the precise measurement of H atomic transition frequencies.

To observe one-photon transitions in the optical region without the Doppler effect, the atoms must be localized in an area smaller than the wavelength of the probe electromagnetic wave. Two methods were developed for the precise measurement of transition frequencies in the optical region. In one of these methods, an ion is trapped in an apparatus (figure 5.5) using an RF-electric field, the intensity of which is proportional to the distance from the trap center. A single ion has a vibrational motion around the trap center, where the electric field is zero. The amplitude of the vibrational motion is reduced to less than the optical wavelength by laser cooling. A fractional uncertainty lower than 10^{-17} was attained with the $^{27}\text{Al}^+$ [7] and the $^{171}\text{Yb}^+$ [8] transition frequencies.

A method was also developed to measure the transition frequencies of neutral atoms (figure 5.6). The atoms are laser cooled and trapped at the anti-node of the standing wave of a laser light (an optical lattice). Measurements are performed using many atoms to achieve low statistical uncertainty with a short measurement time. The trap laser light induces a Stark shift in the transition frequency, which can be positive or negative depending on the wavelength of the trap laser. By choosing the trap laser wavelength—called the magic wavelength—where the Stark shifts in the

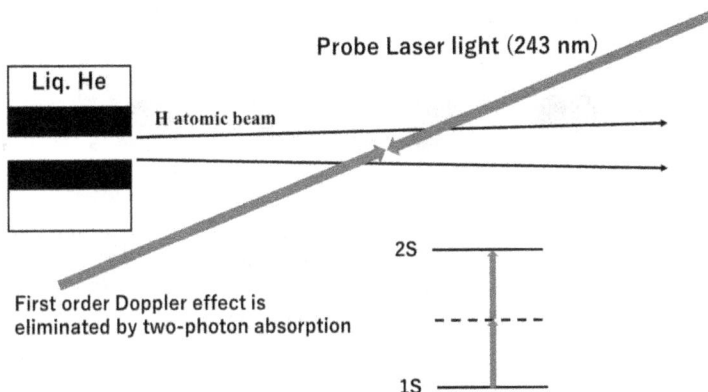

Figure 5.4. Schematic for measuring the H 1s–2s transition frequency by two-photon absorption.

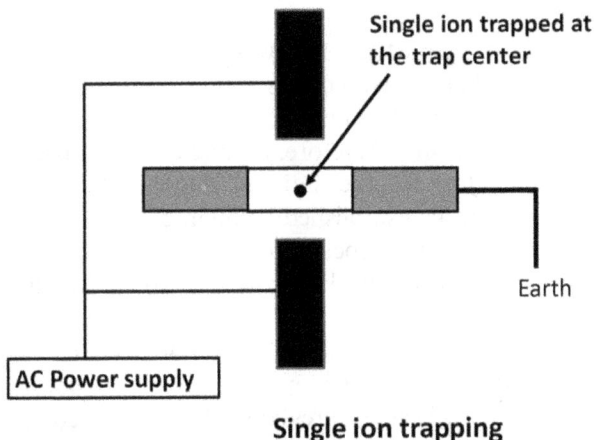

Single ion trapping

Figure 5.5. Schematic of the structure of an ion trap apparatus.

upper and lower states are equal, the shift in the transition frequency is eliminated. We call this type of atomic clock an 'atomic-lattice clock'.

An accuracy on the order of 10^{-18} has been obtained for the ^{87}Sr transition frequency at RIKEN-University of Tokyo (Japan) and the Joint Institute for Laboratory Astrophysics (JILA, USA) [9, 10]. The ^{87}Sr transition frequency has been measured at many other institutes in France, Germany, the US, Japan, the UK, and Italy. The idea was also applied to the transition frequencies of ^{171}Yb and ^{199}Hg atoms. The ratio of the ^{87}Sr and ^{171}Yb transition frequencies is given with a fractional uncertainty of 5×10^{-17} [11].

Measurements with a single trapped ion provide advantages in lowering the systematic uncertainty, because the Coulomb interaction between electrons and nuclei is stronger than that for neutral atoms with the same electron number, and the energy gaps between different states are larger. Lattice-clock measurements with

Atoms are trapped at the anti-node of light standing wave

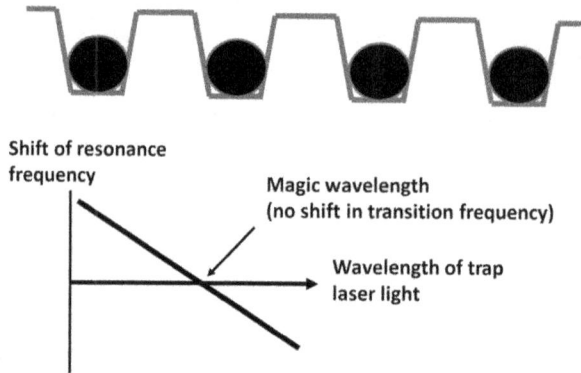

Figure 5.6. Concept behind the atomic-lattice clock. Atoms are trapped at the anti-nodes of a standing wave of a laser light with the frequency set where the transition frequency shift is zero.

many atoms provide more advantages in lowering the statistical uncertainty with a short measurement time than measurements with a single trapped ion.

The continuous operation of atomic clocks with optical atomic transition frequencies is currently difficult. Therefore, atomic clocks in the microwave region are used as source oscillators for the national standard time. However, accurate national standard times can be established through corrections based on measurements obtained using a ^{87}Sr lattice clock taken over one 3 h period per week [12]. In the near future, each national standard time is likely to be established using atomic clocks operating in the optical region.

It seemed opportune to establish a new standard of time and frequency based on an atomic transition frequency in the optical region. However, deciding which atomic transition frequency is best requires some thought. A redefinition of time and frequency is expected sometime after 2026.

5.3 Length standard given by the constant value of the speed of light

Since 1960, the length standard has been given by the wavelength of the light emitted by krypton-86 (^{86}Kr), and its uncertainty was on the order of 10^{-9} [13]. Although this was the first standard given in on a micro scale, the measurement uncertainty was not high enough without using lasers.

Note that the wavelength of light λ is given by $\lambda = c/f$, where c is the speed of light in a vacuum, and f is the frequency. In principle, the uncertainty of the wavelength is the same as that of the frequency, assuming the constant value of c. Here, we present the procedure to confirm the constancy of c.

In the 19th century, the speed of light in a vacuum was expressed by Maxwell's equations, which does not have any dependence on an observer. This was a mystery, because velocity depends on the motion of an observer. If observers A and B move along one direction with velocities V_A and V_B, respectively, B will observe the velocity of A to be $V_A - V_B$. Hence, the notion of 'ether' was proposed as the propagator of light in a vacuum. A different value for the speed of light was expected for observers moving with velocity v within the ether. Michelson and Morley tried to observe this effect in the speed of light that would arise with the Earth's orbital and rotational motion [14]. Using an apparatus known as an interferometer (figure 5.7), light was split by a partial mirror into two waves directed parallel and perpendicular to Earth's motion and reflected by mirrors at distances L_1 and L_2 parallel and perpendicular to the direction of Earth's motion, respectively. The reflected rays of light overlap at the partial mirror again, and light interference should be observed related to the difference in the propagation times (T_1 and T_2). If the Earth's velocity of motion is v, then

$$T_1 = 2L_1/c[1-\left(\frac{v}{c}\right)^2], \tag{5.1}$$

and

$$T_2 = 2L_2/c\sqrt{1-\left(\frac{v}{c}\right)^2}. \tag{5.2}$$

The Earth's velocity of motion changes seasonally with respect to the ether. The change in ($T_1 - T_2$) by the change in (v/c) ($\delta(T_1 - T_2)$) can be detected by the change in the relative phase $\delta\phi = f\,\delta(T_1 - T_2)$. Assuming $L_1 = L_2 = 9$ m, $T_{1,2}$ are 6×10^{-8} s, and it is difficult to measure the change in $T_{1,2}$ with the change in (v/c) by 10^{-4} directly. However, the change in the relative phase is much more sensitive to the change in (v/c). Assuming $f = 4 \times 10^{14}$ Hz, $\delta\phi = 0.1$ was estimated, which was detectable with the accuracy level of the experiment at that time. No change in the

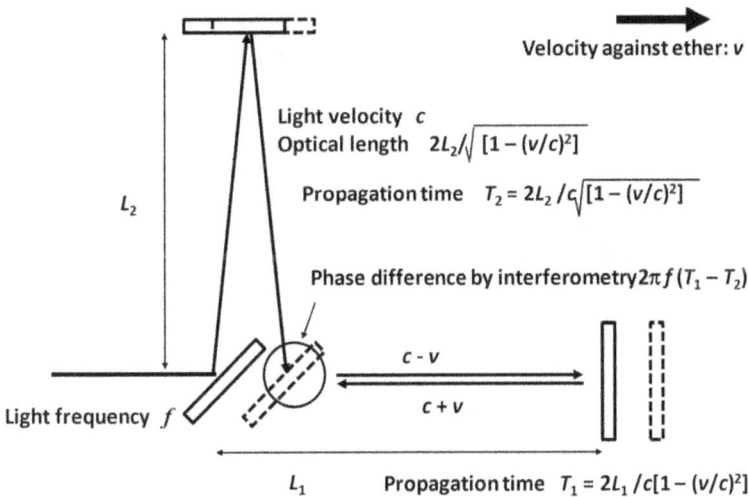

The Michelson-Morley Experiment

Figure 5.7. Experimental apparatus to investigate the influence of the Earth's motion on the speed of light.

speed of light was observed; indeed, the speed of light was concluded to be constant for all observers with an uncertainty of 10^{-8}. The theory of relativity was established based on the constant value of the speed of light, and no phenomena have been discovered to disprove this theory. Therefore, it is reasonable to accept the constancy of the speed of light. To observe any effect in interferometry, using laser light of a single frequency is preferable. With a setup using a laser, Brillet and Hall reduced the experimental uncertainty to 10^{-15} in 1979 [15].

In 1983, the speed of light in a vacuum was defined to be 299 792 458 m s^{-1} without any uncertainty. Hence, the wavelength of a laser light known to have a low frequency uncertainty offers a means to provide an accurate length scale.

The wavelength in the optical region is more useful than that in the microwave region because of the finer scale of length. Now the precise measurement of frequencies in the optical region is possible, which has led to a significant reduction of uncertainty for the length standard.

Now, laser range finders are used to measure the distance to a target from the propagation time of the emitted and reflected pulsed laser light [16]. Its measurement accuracy is limited by the rise and fall time of the laser pulse and the fluctuation of the refractive index (constancy of the speed of light is guaranteed only in a vacuum). The measurement uncertainty is reduced to 10^{-9} by using two-wavelength double-heterodyne interferometry [17].

5.4 Mass standard on a micro scale

The properties of solid materials on a macro scale are not universal and are difficult to use as standards. The standards of time, frequency, and length are now established with the properties of atoms in a gaseous state, whose properties are universal. The mass of atoms and some elementary particles are also universal and have potential for the establishment of a new standard.

The mass ratio between atoms and electrons is measured as follows. When the magnetic field B_z is applied, particles with an electric charge of q and mass m_a have a circular motion called cyclotron motion on the plane parallel to the magnetic field with the frequency of $qB_z/2\pi m_a$. From the ratio of the frequencies of cyclotron motions of X^{j+} ion (mass $m(X^{j+})$) and electron (mass m_e), $m_e/m(X^{j+})$ is obtained. For example, $m_e/m(^{12}C)$ was determined to be $4.571\ 499\ 259 \times 10^{-5}$ from the ratio of the electron frequency and $^{12}C^{6+}$ ion (the mass of a neutral atom is obtained considering the mass of lacking electrons and the binding energy), as shown in figure 5.8 [18]. Therefore, m_e can be the standard of mass on a micro scale for the discussion of the mass of atoms.

How can we determine m_e? It can be obtained from the transition frequencies f of hydrogen atoms changing the number of the main quantum, which are given by

$$\frac{f}{c} = R_\infty\left[\frac{1}{m^2} - \frac{1}{n^2}\right],\tag{5.3}$$

where m and n are integers ($m = 1$: Lyman series, $m = 2$: Balmer series) and R_∞ is the Rydberg constant, which is given by

Magnetic field: B_z

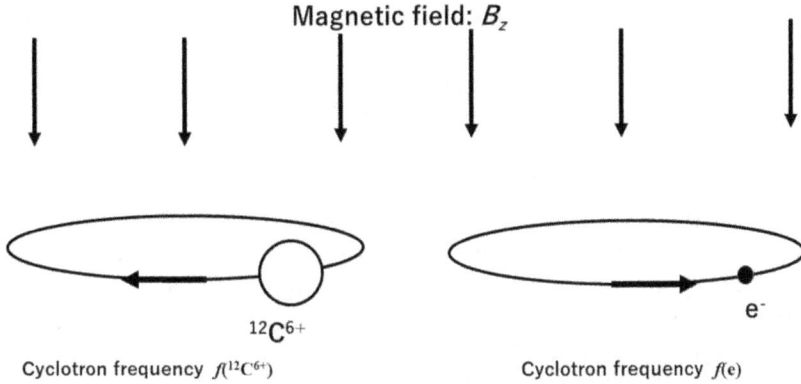

$$m(^{12}C^{6+})/m_e = f(e)/[6f(^{12}C^{6+})]$$

Figure 5.8. Measurement of the mass ratio between $^{12}C^{6+}$ ion and electron e^- comparing the frequencies of cyclotron motion (circular motion of charged particle under a magnetic field).

$$R_\infty = \frac{m_e e^4}{8\varepsilon_0^2 h^3 c},\qquad(5.4)$$

where e is the elemental electric charge, ε_0 is the electric susceptibility, and h is the Planck constant. Using the fine structure given by

$$\alpha = \frac{e^2}{4\pi\varepsilon_0 hc}.\qquad(5.5)$$

Equation (5.4) is rewritten as

$$R_\infty = \frac{\alpha^2 m_e c}{2h}.\qquad(5.6)$$

Then m_e is obtained by

$$m_e = \frac{2R_\infty h}{\alpha^2 c}.\qquad(5.7)$$

The value of R_∞ (= $1.097\ 731\ 6 \times 10^7\ \text{m}^{-1}$) was obtained with the uncertainty of 10^{-11} from the precise measurement of transition frequencies of H and D atoms [19]. The value of α ($7.297\ 352\ 566\ 4 \times 10^{-3}$) was measured with the uncertainty of 2×10^{-10} from the recoil of ^{133}Cs atom in a matter-wave interferometer [20]. The uncertainty of m_e is dominated by that of the Planck constant ($6.626\ 070\ 04 \times 10^{-34}$ kg m^2 s^{-1}), which is 1.2×10^{-8} [21]. In November 2018, it was decided that the Planck constant should be defined as a value without uncertainty starting in May 2019, and m_e can be obtained as a value with an uncertainty on the order of 10^{-10}.

The dimensions of the Planck constant are denoted by M L^2 T^{-1} (M: mass, L: length, and T: time). Defining the speed of light c as a value without uncertainty,

$L = c\mathrm{T}$, and the uncertainty of the length is given by that of time and frequency. Defining the Planck constant as a value without uncertainty, $\mathrm{M} = (h/c^2)\,\mathrm{T}^{-1}$, and the uncertainty of mass is also given by the uncertainty of time and frequency. It corresponds to the fact that the uncertainty of m_e is given by the uncertainties of the Rydberg constant and the fine structure constant, measured by the precise measurement of atomic transition frequencies using lasers.

The mass standard on a macro scale can also be defined with higher accuracy if we can estimate the number of atoms. A mass of 1 kg can be obtained from a 93.6 mm-diameter sphere of silicon (^{28}Si) atoms (figure 5.9). ^{28}Si was chosen because ultra-pure defect-free monocrystalline silicon already exists in fabricated form.

The natural abundance of ^{28}Si isotope is 0.922, which can be improved up to 0.9999 by isotope separation. The lattice defect is minimized by using a sphere. The lattice spacing between atoms in the crystal $d = 0.192$ nm (1.92×10^{-7} mm) can be measured by x-ray diffraction analysis. With this method, the x-ray diffraction is intensive with the x-ray angle θ satisfying $2d = \lambda_X \sin \theta$, where λ_X is the x-ray wavelength, and the measurement uncertainty of d dominated by that of the x-ray wavelength is significant. The measurement uncertainty of d was reduced by the x-ray interferometry (figure 5.10) [22]. With this method, x-ray irradiation is applied to crystal 1. Transparent and diffracted x-rays are reflected by a mirror and overlap crystal 2. The position of crystal 2 is scanned, and the intensity of the x-rays passing through crystal 2 changes with the periodicity of d. The position change of crystal 2 is monitored by laser interferometry. The accurate value of the x-ray wavelength is not required with this method, and the measurement uncertainty was reduced drastically.

The size of this Si sphere is measured using optical interferometry to an uncertainty of 0.3 nm (3×10^{-7} mm)—roughly the size of a single atom [23]. With this method, the mass uncertainty of the Si sphere is expected to be reduced compared to that of the kg standard mentioned in section 3.5. In November 2018, it was decided that a new mass standard should be used starting in May 2019. A detailed explanation is provided in [24]. The development of laser light sources made it possible to measure the size of a ^{28}Si crystal with low uncertainty. This

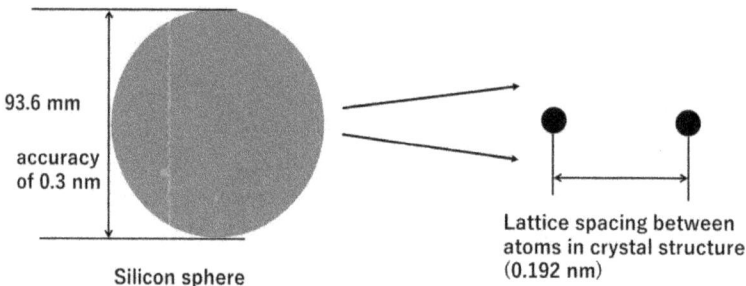

93.6 mm

accuracy of 0.3 nm

Silicon sphere

Lattice spacing between atoms in crystal structure (0.192 nm)

We can measure the number of atoms
(2.15254×10^{25})

Figure 5.9. Schematic of the future mass standard, a pure Si sphere.

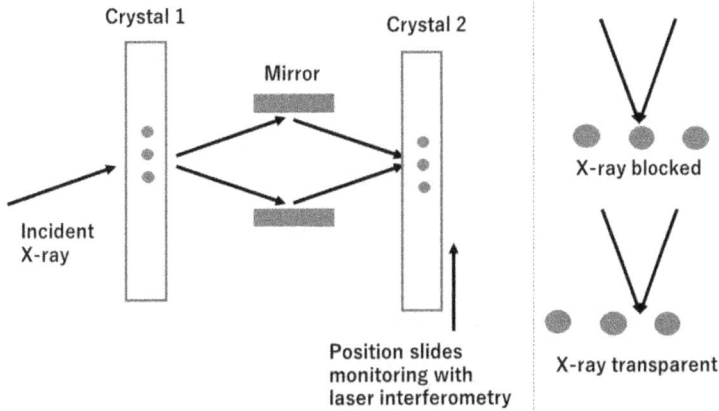

Figure 5.10. Measurement of lattice spacing between atoms in crystal structure.

measurement should be performed at the same temperature as the measurement of the lattice spacing between atoms. Therefore, the temperature should be stabilized within 0.001 K.

5.5 Determination of the Avogadro constant

The Avogadro constant has been defined as the number of ^{12}C atoms contained in 12 g (0.012 kg) of carbon. The mol electron mass is given by $A_e = m_e/m(^{12}\text{C}) \times 0.012$ kg. Then the Avogadro constant is given by

$$N_A = \frac{A_e}{m_e} = \frac{A_e \alpha^2 c}{2 R_\infty h}.$$ (5.8)

The uncertainty of N_A is dominated by that of the Planck constant because the uncertainty of $A_e \alpha^2 c/R_\infty$ is 7×10^{-10} [25]. Therefore, the definition of the Planck constant as a value without uncertainty is equivalent to the definition of the Avogadro constant [24].

5.6 Definition of elementary charge and new definition of electric current

In principle, the electric current of 1 A was defined as the flow of 1 C s^{-1}. However, the measurement of the flow of charged particles (mainly electrons) was not realistic, and electric current has been measured from the Lorenz force (the force induced by the magnetic field given by an electric current). Nowadays, microchip electric circuits have been developed, with which the number of flowing electrons is countable. If the elementary charge e (absolute value of charge of electrons and protons) is known, an electric current can be measured as in principle 1 C s^{-1}.

The elementary charge was first measured by Millikan at 1909 (published in 1911 [26]) to be 1.592×10^{-19} C. Its recommended value is now $1.602\,176\,620\,8 \times 10^{-19}$ C. Since then, the values of e^2/h and e/h have been measured with low uncertainties using the quantum Hall effect [27] and the Josephson effect [28], respectively. Therefore, the definition of the Planck constant is equivalent to the definition of the elementary charge.

It was decided that the value of elementary charge e should be defined as a value without uncertainty starting in May 2019. Then an electric current can be measured from the number of flowing charged particles with the same uncertainty as that of time and frequency (figure 5.11).

5.7 Temperature measurement with defined Boltzmann constant

The thermodynamic temperature T has been defined by the triple point of water as 273.16 K. The uncertainty of this definition is 0.1 mK, which was estimated using distilled water with defined abundance ratios of various isotopes. It is not realistic to use water as defined, and the uncertainty for actual temperature measurement is higher.

The Boltzmann constant k_B ($= 1.380\,648\,52 \times 10^{-23}$ kg m^2 s^{-2} K^{-1}) is a coefficient to give the parameter of the energy distribution $k_B T$. In November 2018, it was decided that the Boltzmann constant k_B should be defined as a constant without uncertainty, and a new temperature standard will be used starting in May 2019. With this new definition, the thermodynamical temperature is a parameter of energy distribution. For example, the temperature of a high-temperature object can be determined by precise measurement of the peak frequency of the blackbody radiation (radiation of electromagnetic waves from objects; maximum frequency is proportional to T and total power is proportional to T^4).

The stabilization of temperature is required at a high level, for example, for an optical cavity for which the laser frequency is stabilized.

5.8 Measurement uncertainty of luminal intensity

The luminal intensity of one candela is defined as the brightness of light with a frequency of 540 THz (wavelength 555 nm) with a radiant intensity of 1/683 W per steradian (section 3.8). This definition has not changed since the definition of the

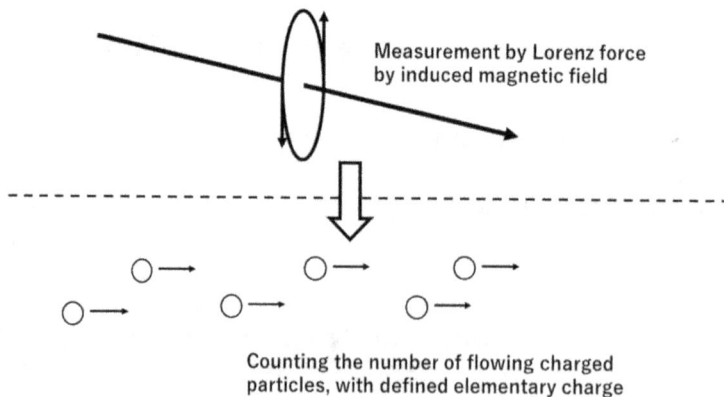

Figure 5.11. Measurement of an electric current before and after the definition of elementary charge.

Planck constant, the Boltzmann constant, and the elementary charge. However, the energy of one photon with a given frequency f is given by hf, whose uncertainty is given by that of the frequency after the definition of the Planck constant. The radiant intensity can be measured as the number of flowing photons. Luminal intensity is the frequency weighted power, and the weighting parameter is artificially defined without uncertainty. Therefore, the measurement uncertainty of the luminal intensity became higher after the definition of the Planck constant.

5.9 The role of lasers in the redefinition of physical values

The role of lasers in the development of new definitions of time and frequency, length, and mass is shown in figure 5.12. Since the invention of the laser, the measurement uncertainty of atomic transition frequency has been reduced drastically. A new definition of time and frequency is expected to be introduced after 2026. Using lasers, the constancy of the speed of light was confirmed with the uncertainty of 10^{-15} [15], which made it possible to define the speed of light as a value without uncertainty. The length standard is now given by the wavelength of light with low frequency uncertainty.

The precise measurement of atomic transition frequencies made it possible to reduce the uncertainties of the Rydberg constant, fine structure constant, and so forth. The mass of an electron is obtained by defining the Planck constant as a value without uncertainty. Then the atomic mass can be obtained with low uncertainty. The number of ^{28}Si atoms in a sphere can be measured with the uncertainty of 3×10^{-9}, and the uncertainty of mass on a macro scale can be reduced. With this method, the Avogadro constant can also be defined.

Defining the Boltzmann constant and the elemental electric charge, new definitions of thermomechanical temperature and electric current will be given. The definition of luminal intensity will not be changed, but its uncertainty will be reduced

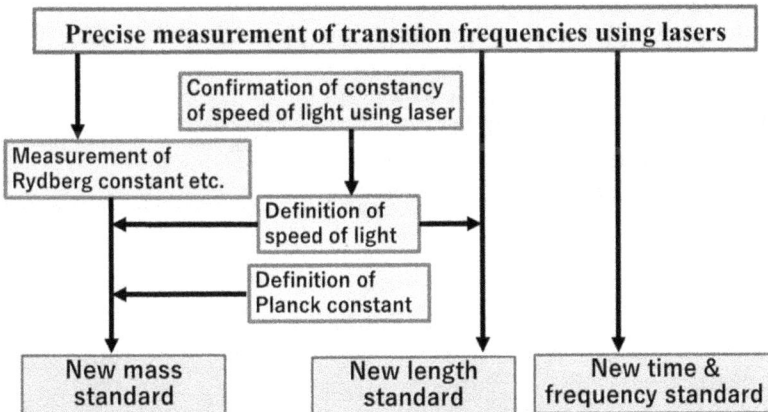

Figure 5.12. Role of precise measurement of atomic transition frequencies using lasers in the development of new definitions of time and frequency, length, and mass.

by defining the Planck constant because the uncertainty of photon energy is given by that of frequency.

Thus, the history summarized in this chapter clearly shows that the invention of the laser led to a revolution that enabled the reduction of the measurement uncertainties for all physical values.

References

[1] Hasegawa A *et al* 2004 *Metrologia* **41** 257
[2] Clairon A, Salomon C, Guellati and Phillips W 1991 *Europhys. Lett.* **16** 165
[3] Heavner T P *et al* 2005 *Metrologia* **42** 411
[4] Ovchinnikov Y and Marra G 2011 *Metrologia* **48** 87
[5] Knappe S *et al* 2005 *Opt. Lett.* **30** 2351
[6] Niering M *et al* 2000 *Phys. Rev. Lett.* **84** 5496
[7] Chou C W *et al* 2010 *Phys. Rev. Lett.* **104** 070802
[8] Huntemann N *et al* 2016 *Phys. Rev. Lett.* **116** 063001
[9] Ushijima I *et al* 2015 *Nat. Photon.* **9** 185
[10] Nicholson T L *et al* 2015 *Nat. Commun.* **6** 6896
[11] Nemitz N *et al* 2016 *Nat. Photon.* **10** 258
[12] Hachisu H *et al* 2018 *Sci. Rep.* **8** 4243
[13] Baird K M and Howlett L E 1963 *Appl. Opt.* **2** 455
[14] Michelson A A 1881 *Am. J. Sci.* **22** 120
[15] Brillet A and Hall J L 1979 *Phys. Rev. Lett.* **42** 549
[16] https://theopticworld.com/accurate-laser-rangefinders/
[17] Kuramoto Y and Okuda H 2013 *Talk presented at the Int. Workshop on future Linear Colliders (LCWS13, Tokyo)*
[18] Farnham D L *et al* 1995 *Phys. Rev. Lett.* **75** 3598
[19] Pohl R *et al* 2017 *Metrologia* **54** L1
[20] Parker R H *et al* 2018 *Science* **360** 191
[21] https://physics.nist.gov/cgi-bin/cuu/Value?h
[22] Bonse U and Hart M 1965 *Z. Phys.* **188** 154
[23] Bartl G *et al* 2011 *Metrologia* **48** S96
[24] Schlamminger S 2018 *Redefining the Kilogram and Other SI Inits* (Bristol: IOP Science) pp 18–22
[25] Mohr P J *et al* 2012 *Rev. Mod. Phys.* **84-4** 1527
[26] Millikan R A 1911 *Phys. Rev.* **32** 349
[27] Klitzing K V *et al* 1980 *Phys. Rev. Lett.* **45** 494
[28] Josephson B D 1962 *Phys. Lett.* **1** 251

IOP Publishing

Measurement, Uncertainty and Lasers

Masatoshi Kajita

Chapter 6

Measurement uncertainties and physics

6.1 Introduction

Physics concerns the laws of nature, from which we can make predictions and retrodictions regarding the behavior of phenomena. It is based on measurement results. For example, with the measurement results of $(x, y) = (1, 1), (2, 2), (3, 3), (5, 5)$, the law of $y = x$ will be established. However, measurement results always have uncertainties. If there are measurement uncertainties of 10%, we cannot distinguish the difference between $y = x$ and $y = 10 \sin(x/10)$. If we give the law $y = x$, there may be some discrepancy when a measurement result with $x = 20$ is obtained (figure 6.1).

Physical laws have been established based on the measurement results that have been obtained up to a given time. With the discovery of new phenomena, new discrepancies with the previous physical laws have been discovered, and new physical laws have been developed.

New phenomena have been discovered through the reduction of measurement uncertainties. If the uncertainty of measurements was reduced to 0.5% and $(x, y) = (1, 0.99), (2, 1.98), (3, 2.95), (5, 4.79)$ were obtained, $y = 10 \sin(x/10)$ becomes more reasonable than $y = x$.

The development of physics is particularly closely correlated to the development of accurate clocks. This is because the reduction of measurement uncertainty has been particularly significant with time and frequency: their uncertainty was much larger than that for length and mass in ancient times, but time and frequency are now the physical values that can be measured with the lowest uncertainty.

Section 6.2 reviews the correlation between the development of physics and clocks. Several mysteries of modern physics are then presented, which are expected to be solved by the discovery of slight effects after the further reduction of measurement uncertainties.

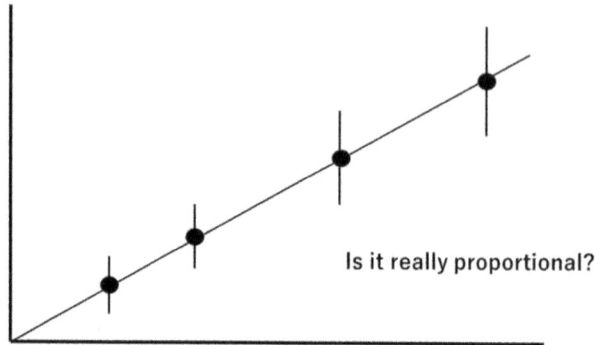

Is it really proportional?

Figure 6.1. All measurements have uncertainty; therefore, laws obtained from measurement results are always questionable.

6.2 History of physics and clocks

Research into natural philosophy started in ancient Greece. Aristotle argued that the Earth was spherical because, going south, constellations along the southern horizon are seen at higher positions, and in a lunar eclipse, we observe a circular shadow of the Earth. However, the Earth was believed to be the center of the Universe, about which the Sun and the stars were believed to revolve. At that time, the time of day was measured with sundials or water clocks with an uncertainty on the order of 1 h, and no discrepancy between observations and the Ptolemaic theory was discernible.

After the invention of the mechanical clock, the accuracy of time keeping was improved. Then, irregular motions of planets among constellations were discovered, which was a serious discrepancy from the Ptolemaic theory. In 1510, Copernicus published his idea that the Earth revolved around the Sun, which was called Copernican theory. It was not accepted at that time because he assumed that orbits were circular and his new theory could not account for the planetary observations in such detail.

The accuracy of time measurement was drastically improved by the discovery of the periodicity of the pendulum's swing by Galileo. In 1656, the pendulum clock was invented by Huygens. At that time, the law of the motion of the planets around the Sun was established by Kepler, and Copernican theory was mostly established. There was still an argument that the revolution motion of the Earth must cause the periodical change in the directions the stars that were observed during one year (called parallax of the year). The parallax of the year could not be observed with the measurement uncertainty at that time. After the reduction of the measurement uncertainty in direction, the parallax of the year was observed with many stars, and the distances to the stars were determined to be much greater than previously thought (therefore, too small a change of direction to be observed in the 17th century).

The law of the motion of planets around the Sun and the objects on the Earth were summarized by Newtonian mechanics. This was roughly during the same period when the pendulum clock was invented. This coincidence portends the beginnings of the close correlation between the developments of clock technology and physics.

Until the middle of the 17th century, the speed of light was believed to be infinite. This is because no phenomenon was observed that highlighted the finite propagation time of light with the measurement accuracy of time. However, this supposition of the propagation of light was no longer tenable after the invention of the pendulum clock. The period taken by Io, one of Jupiter's satellites discovered by Galileo in 1609, to complete an orbit was not constant. In 1676, Rømer considered a finite propagation time for light between Jupiter and the Earth. The distance between the Earth and Jupiter changes as both planets move around the Sun. Hence, the propagation time of light varies. Based on this idea, the speed of light was estimated to be $c = 2.2 \times 10^8$ m s^{-1}, which is 26% lower than the present value [1].

The accuracy of the speed of light was improved through the measurements obtained by Fizeau, Foucault, and others [2, 3]. Figure 6.2 shows the principle of the measurements taken by Fizeau. Light passing through a gear was reflected by a mirror 9 km away. The speed of light was determined from the velocity of rotation of the gear when reflected light was blocked. The principle of Foucault's measurement was the same as that used in Fizeau's measurement except that a rotating mirror was used instead of a gear. Foucault's result was 2.98×10^8 m s^{-1}, which is 0.6% different from the present value.

In the 18th and 19th centuries, several laws of electricity and magnetism were discovered, and they were unified in a set of equations known as Maxwell's equations. These equations indicate that the changes in electromagnetic fields propagate as a wave. The speed of propagation in a vacuum, as derived by Maxwell's equations, was in good agreement with the speed of light obtained by

Measurement of light velocity by Fizeau

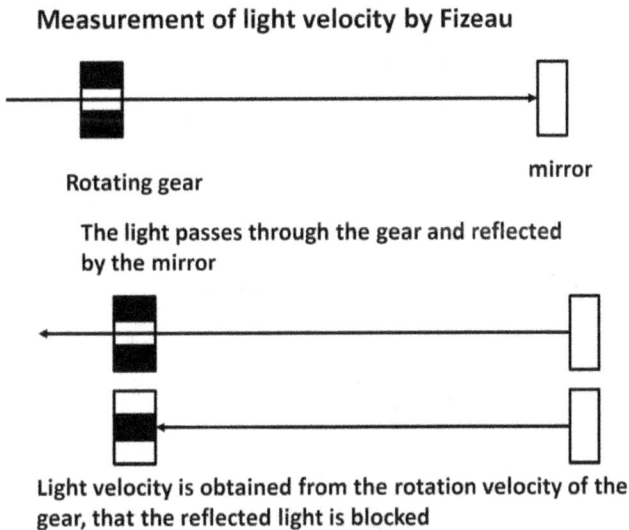

Rotating gear   mirror

The light passes through the gear and reflected by the mirror

Light velocity is obtained from the rotation velocity of the gear, that the reflected light is blocked

Figure 6.2. Principle involved in the measurement of the speed of light by Fizeau.

Fizeau and Foucault. Maxwell stumbled upon the idea that light is an electromagnetic wave. This idea was experimentally confirmed by Hertz in 1887.

Maxwell's equations indicate that the speed of light in a vacuum is constant for all observers with any motion (generally, the measurement of velocity depends on the motion velocities of observers). The notion of 'ether' was proposed as the propagator of light in a vacuum. It was expected that the difference for the speed of light would be induced by the motion velocity within the ether. Michelson and Morley tried to observe this effect in the speed of light that would arise with Earth's orbital and rotational motion, but the result of this experiment indicated the constancy of the speed of light with an uncertainty of 10^{-8} [4] (section 5.3). The constancy of the speed of light was later confirmed with an uncertainty of 10^{-15} [5].

6.3 Confirmation of relativistic effects

The theory of special relativity was established based on the constancy of the speed of light. According to this theory, time goes slower in a frame moving with velocity v with a factor of $\sqrt{1-(v/c)^2}$ (figure 6.3) [6]. With the velocity of 300 m s^{-1} (1080 km h^{-1}), the difference of time lapse is 5×10^{-13}, which is difficult to detect with the measurement uncertainty of quartz clocks. However, this effect is significant with measurements obtained using atomic clocks, based on the transition frequencies of atoms.

With this relativistic effect, the transition frequency observed using atoms moving with velocity of v is lower than the real transition frequency with a factor of $\sqrt{1-(v/c)^2}$ (called the quadratic Doppler shift). With Cs atomic clocks using atomic beams, quadratic Doppler shifts depending on the atomic velocity distribution can be observed [7]. The quadratic Doppler shift is observed clearly from a comparison of Cs atomic clocks with an atomic beam and an atomic fountain (quadratic Doppler shift $< 10^{-17}$). The quadratic Doppler shift induced by velocities in the

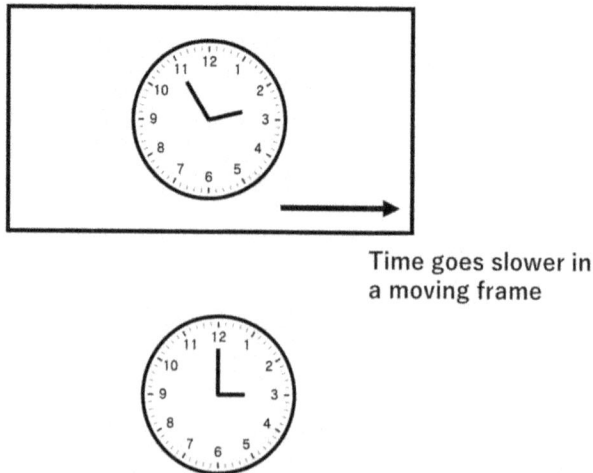

Time goes slower in
a moving frame

Figure 6.3. According to the theory of special relativity, time goes slower in a moving frame.

range 0 to 40 m s^{-1} was measured using the ^{27}Al$^+$ transition frequency, the fractional uncertainty being on the order of 10^{-18} [8].

The theory of general relativity is based on the equivalency principle between gravitational force and inertial force [6]. According to this theory an optical path is bent by gravitational force, which was confirmed in 1919 by Eddington when observing a distant star during a solar eclipse [9]. From the bent optical path and the constant speed of light, the propagation time of light becomes longer. Therefore, time goes slower with the gravitational potential of $m\Phi$ (m: mass) with a factor of $\sqrt{1-(2\Phi/c^2)}$ (figure 6.4).

Therefore, the observed atomic transition frequency is lower in a gravitational potential (gravitational red shift). This effect is observed as a change in the transition frequency with changing altitude (10^{-16} m^{-1}). It was measured by comparing the frequencies of Cs atomic clocks located in a valley and on a nearby mountain [10]. The change in the gravitational red shift with the difference of altitude of 15 cm was observed using the ^{27}Al$^+$ transition frequency [8].

The bending of an optical path is a distortion of space, which propagates as a wave (called a gravitational wave) with the speed of light. We can imagine a model with an object moving on a flexing mat (figure 6.5). For gravitational waves, the size in the direction perpendicular to the propagation repeatedly expands and contracts. Its effect has been very difficult to observe, because the change in size is so small, being 10^{-21}, corresponding to observing a change in the size of an atom over the distance between the Sun and the Earth.

The existence of gravitational waves was indirectly confirmed from the decrease in the orbital period of a binary neutron star PSR 1913 + 16 (7.75 h) by 7.65×10^{-5} s yr^{-1} [11] (figure 6.6). This result suggests that the energy loss is in the form of radiating gravitational waves. The measured variation in the orbital period is in good agreement with the estimation obtained from the theory of general relativity to within 0.2%. The accuracy of the atomic clock contributed to measuring the orbital period with high precision (present accuracy of 1.5×10^{-5} s).

The direct detection of gravitational waves was achieved by laser interferometry. The apparatus (figure 6.7) consists of a laser (Nd:YAG laser, wavelength: $\lambda = 1064$

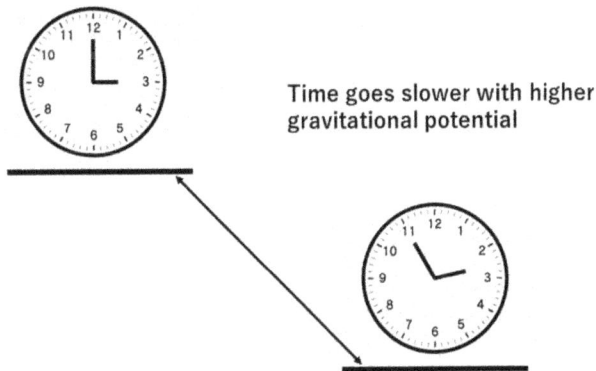

Time goes slower with higher gravitational potential

Figure 6.4. According to the theory of special relativity, time goes slower with gravitational potential.

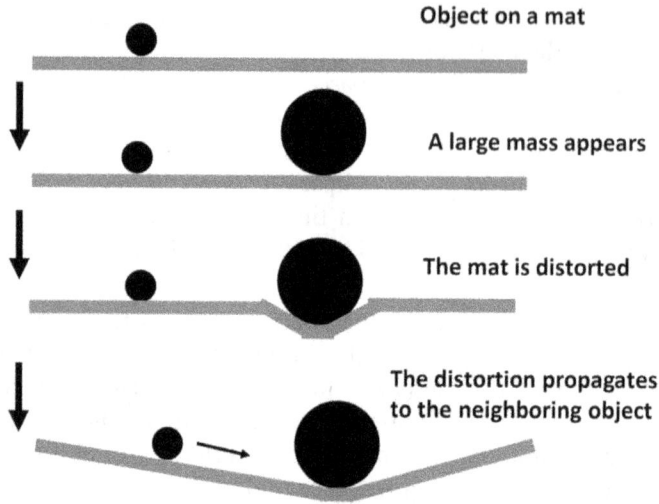

Figure 6.5. Space distortion induced by gravity can be imagined with the distortion of the surface of a flexing mat when a large object is placed on it.

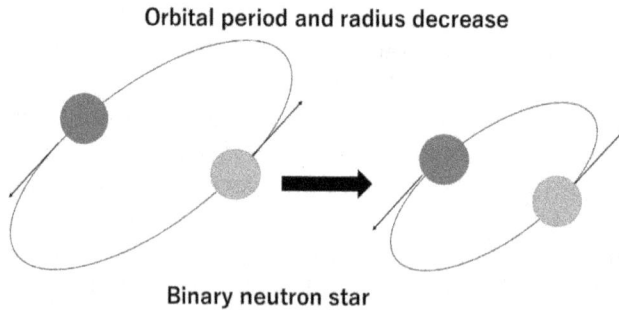

Figure 6.6. Orbital period and radius of a binary neutron star decrease because energy is lost with the generation of a gravitational wave.

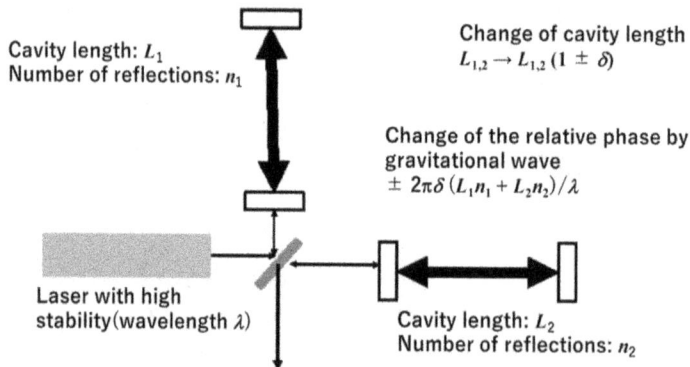

Figure 6.7. Laser interferometry system to detect gravitational waves.

nm) that shines light onto a half mirror, which splits and redirects the light into two perpendicular directions. In the arms of the interferometer (resonators of length $L_{1,2}$), the light is reflected many times ($n_{1,2}$ times). Both beams are brought together in the half mirror again to form an interference pattern exhibiting a phase difference of $\phi = 2\pi(n_1 L_1 - n_2 L_2)/\lambda$. When the gravitational wave changes the cavity length by $L_1 \rightarrow L_1(1 \pm \delta)$ and $L_2 \rightarrow L_2(1 \mp \delta)$, a change in the phase difference of $\delta\phi = 2\pi\delta$ $(n_1 L_1 + n_2 L_2)/\lambda$ is observed. The frequency stability (10^{-15}) of the laser light was important in attaining high sensitivity to slight changes in cavity length. The laser power also must be stabilized to 10^{-6} because fluctuations in radiation pressure change the cavity length.

The first direct observation of a gravitational wave was made on 14 September 2015 at the Laser Interferometer Gravitational-wave Observatory (LIGO) in Hanford (Washington, US) and Livingston (Louisiana, US) [12]. Precision measurements regarding the slight difference in detection times at both laboratories are important in finding the source of the gravitational wave. The detected waveform (frequency 30–250 Hz) matched the prediction of general relativity for a gravitational wave emerging from the merger of a pair of black holes of 36 and 29 solar masses at a distance of 1.3 billion light years.

The relativistic effects mentioned in this subsection are very slight, so they were detected only with ultra-precise measurement of time and frequency by atomic clocks (including frequency-stabilized lasers to detect gravitational waves). There have been no experimental results to contradict the theory of relativity. However, we cannot conclude that it is a perfect theory. We cannot guarantee the applicability of the theory of general relativity with an ultra-high gravitational potential. Unification of the theory of general relativity and quantum mechanics has never been attained, and there may be some requirement for new physics in the future.

6.4 Symmetry between particles and antiparticles

In 1928, Dirac derived an equation to obtain the quantum mechanical waveform of the electron (mass of m_e) based on the theory of special relativity [13]. The Dirac equation is a 4×4 matrix equation for a wavefunction given as a four-dimensional column vector. The four solutions obtained correspond to spin $\pm 1/2$ and invariant-mass energies $\pm m_e c^2$. The existence of positrons with a positive charge and a mass of m_e was expected from the solution of the negative mass. This idea was confirmed by the discovery of positrons, and this was the first discovery of antiparticles. Antiprotons and antineutrons were also discovered later. For all particles, antiparticles also exist.

From our perspective, we see only matter constructed from particles. It is a mystery of modern physics that we do not see antiparticles. It is thought that there were particles and antiparticles at the dawn of the Universe. The antiparticles may have disappeared by pair annihilation with particles. It is thought that the initial number of particles was slightly larger than that of antiparticles (ratio of $1 + 10^{-9}$: 1); therefore, particles still remain in the Universe after the disappearance of antiparticles. The mystery of the asymmetry between the numbers of particles and antiparticles has not yet been solved.

What is the relationship between a particle and an antiparticle? Is it just the conjugation of the electric charge? This is not correct, because the spin of a neutrino is only −1/2, and it is only +1/2 for an antineutrino. Antiparticles were imagined to be particles after the charge conjugation and mirror reflection (CP-symmetry), as shown in figure 6.8.

Violation of the CP-symmetry was discovered in 1964 from the transformation between different neutral kaons having different CP-symmetry [14]. The Kobayashi–Maskawa theory shows that the violation of CP-symmetry is possible if there are six quarks [15]; the existence of all six quarks was experimentally confirmed. This particle predominance in the Universe can be explained by the violation of CP-symmetry, but it is not enough with the estimation by the Kobayashi–Maskawa theory.

In 1998, violation of the time reversal symmetry (T-symmetry) was discovered [16]. Transform between a particle and an antiparticle is possible, but the rate of the transformation from an antiparticle to a particle is higher than that from a particle to an antiparticle by 0.6% (four times larger than the uncertainty), as shown in figure 6.9. This result indicates that the number of particles is larger than the number of antiparticles.

A fundamental idea in modern physics is that antiparticles must have the image of particles after charge conjugation, mirror reflection, and time reversal (CPT-symmetry). For kaon decay, CPT-symmetry is conserved [16]. The equality of mass as well as the absolute value of the electric charge between particles and antiparticles is required for CPT-symmetry. The equality of |charge|/mass between a proton and an antiproton was confirmed by comparing the cyclotron frequencies set by a magnetic field of 1.95 T, and an accuracy of 7×10^{-11} was achieved [17].

The equality of the 1s–2s transition frequencies (transition from the ground state to the first excited state with the frequency of 2466 THz) between an H atom and an anti-H atom (consisting of an antiproton and positron) was also confirmed (figure 6.10) with an uncertainty of 2×10^{-12} [18]. For the precise measurement of the transition frequency of anti-H atoms, long observation time was required. Using an inhomogeneous magnetic field, anti-H atoms were trapped for up to 1000 s [19].

$\pm e$ $\mp e$

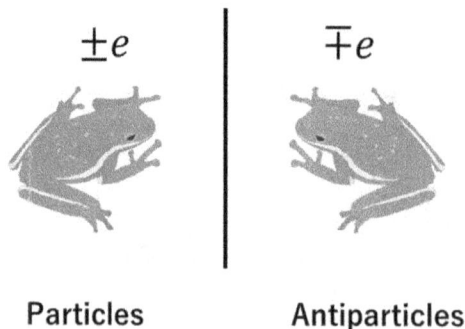

Particles   **Antiparticles**

Figure 6.8. Image of CP-symmetry.

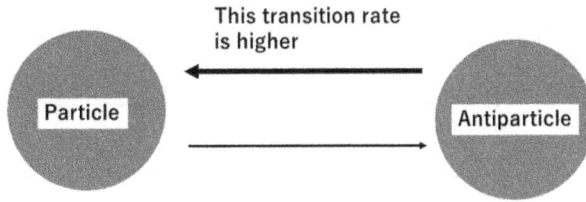

Violation of T-symmetry with kaon

Figure 6.9. Violation of the time reversal symmetry with the transformation between a particle and an antiparticle of kaon. The rate of transform from an antiparticle to a particle is higher than that from a particle to an antiparticle.

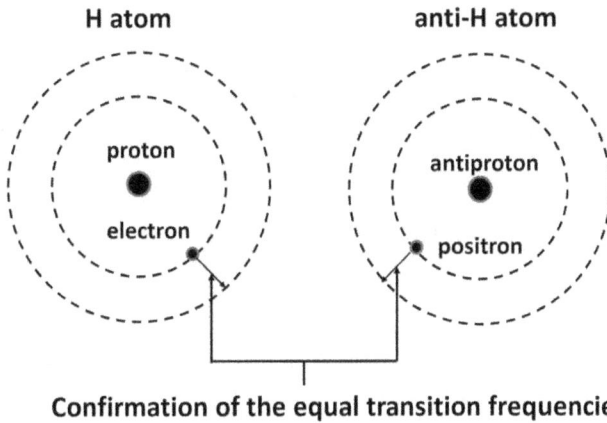

Confirmation of the equal transition frequencies

Figure 6.10. Comparison of transition frequencies of an H atom and an anti-H atom.

Precise measurement of transition frequencies played an important role in the discovery of the violation of CP-symmetry and the confirmation of CPT-symmetry. Nevertheless, there is still the possibility that CPT-symmetry violation can be detected with smaller measurement uncertainties.

6.5 Observation of the vacuum energy (quantum electrodynamics)

Taking the relativistic effect into account, the hydrogen energy structure can be solved by the Dirac equation. However, there are still discrepancies with experimental results. The first problem is that there is a slight energy gap (transition frequency 1.04 MHz) between two states ($2^1S_{1/2}$ and $2^1P_{1/2}$ states), which should have exactly the same energy level according to the Dirac equation [20], as shown in figure 6.11. This energy gap is called the Lamb shift. This energy gap was attributed to the interaction between the vacuum energy fluctuations and the hydrogen electron in different orbits ($2^1S_{1/2}$: spherical symmetric, $2^1P_{1/2}$: polarized). The development of this slight energy gap played a significant role in the development of the field of quantum electrodynamics in light of the vacuum energy fluctuation.

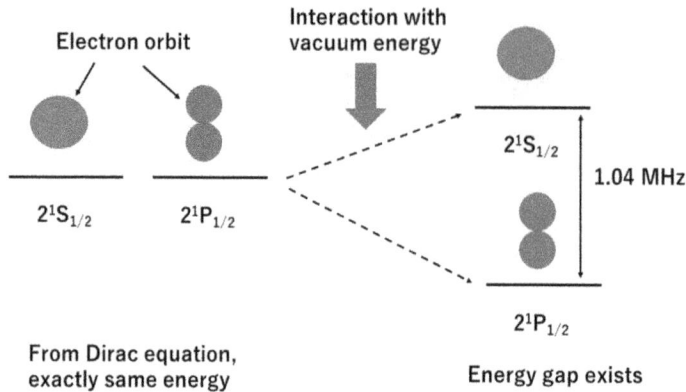

Figure 6.11. Energy gap between two energy states ($2^1S_{1/2}$ and $^2P_{1/2}$) of H atom by Lamb shift.

Quantum electrodynamics provided an explanation for another phenomenon. The coefficient of the interaction force between an electron and a magnetic field (Zeeman coefficient) is 0.1% higher than the estimation from the Dirac equation. The Zeeman coefficient was measured with the uncertainty of 10^{-13} [21], which has been explained within quantum electrodynamics [22].

The precise measurement of the transition frequencies made it possible to discover the phenomena induced by the vacuum energy. Currently no discrepancies between measurement and calculation over the measurement uncertainties have been reported. However, researchers are eager to reduce uncertainties, and it is expected that new discrepancies may be found that will require new physics.

6.6 Proton size puzzle

With quantum electrodynamics, the energy levels of the hydrogen atom can be calculated with high accuracy. The remaining source of discrepancies between experimental and calculation results arises from the finite size of protons. With the measurement of H atoms, it was estimated to be 0.8775 ± 0.005 fm [23]. However, the result obtained with muonic hydrogen atoms (proton + muon) was 0.842 ± 0.001 fm [24] (figure 6.12). The discrepancy in the results is five times larger than the measurement uncertainty; thus, it has become a mystery of modern physics called the 'proton radius puzzle'. Resolving this puzzle may provide a chance to develop new physics.

6.7 Symmetry violation of chiral molecules

Chiral molecules are non-superimposable on their mirror images. The mirror images of a chiral molecules are called optical isomers (figure 6.13). Individual optical isomers are designated as being either right- or left-handed.

The mystery is that the abundances of right- and left-handed optical isomers are not always equal, indicating some violation of chiral symmetry. The reason is believed to stem from the symmetry violation of the mirror reflection induced by a

New Physics?

↑
→ Which is right? ←

0.8775 fm
Hydrogen

0.842 fm
Muonic Hydrogen

Proton radius

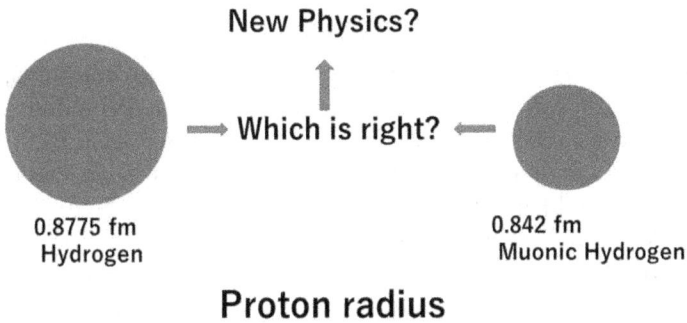

Figure 6.12. Proton radius puzzle.

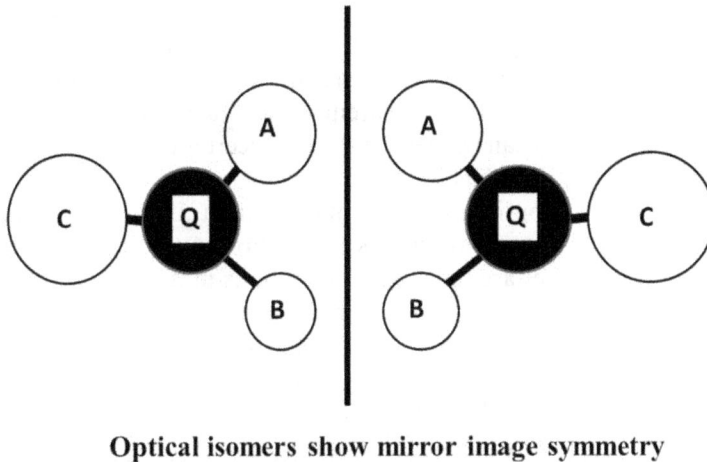

Optical isomers show mirror image symmetry

Figure 6.13. Optical isomers of chiral molecules.

weak nuclear force [25], and there is some difference between the energy structures of two optical isomers. The weak nuclear force works only within the size of the nucleus. But electrons in the S-state (no centrifugal force) can get weak nuclear force because of the non-zero distribution at the position of nuclei.

To confirm this assumption, the slight difference in the transition frequencies must be detectable. The localized abundance of optical isomers is consistent with the theoretical estimate if there is a difference in the string coefficients of atomic bonding with the ratio of 10^{-14}. This accuracy is not easily attained with the transition frequencies of polyatomic molecules at room temperature. Several groups are in the process of developing experimental devices for this purpose [26].

The precision measurement of the molecular transition frequency advances not only physics but also molecular biology because many biologically active molecules are chiral, including amino acids and sugars. Amino acids are mostly left-handed, and sugars are mostly right-handed. With an equal abundance of optical isomers, the appearance of organisms might have occurred differently.

6.8 Search for variations in fundamental constants

Physical laws have been established with many fundamental constants, such as the speed of light c, and the unit of elementary electric charge e. With different values of the fundamental constants, the appearance of the Universe would be quite different. If the ratio of the electromagnetic force to the strong nuclear force was higher, atoms with heavy nuclei could not exist because of the repulsive force between protons (figure 6.14). With smaller electromagnetic force, molecular bonding would not be possible. We would not exist with any other ratio between the four kinds of forces (strong nuclear force, electromagnetic force, weak nuclear force, and gravitational force). The combination of suitable fundamental constants is another of the mysteries of physics. If fundamental constants have a dependence on time and position, we may understand that we are living in an epoch with suitable combinations of fundamental constants; that they arose by chance is not required. In 1937, Dirac mentioned for the first time the possibility (not a requirement) of time-varying fundamental constants [27]. However, the change in variation was expected to be too small to detect with the measurement accuracies at that time. Currently, some transition frequencies can be measured with uncertainties lower than 10^{-16}, and the search for such variations has become an active topic of investigation.

When there is a variation in a fundamental constant X, the energy structures of atoms and molecules change, and there is a variation in the transition frequencies. Some transition frequencies are sensitive to variations in X, while some others have very little sensitivity. The variation in X is monitored from the variation in the ratio between transition frequencies with different sensitivities.

Here, we consider the relationship between the variation in the fine structure constant $\alpha = e^2/(4\pi\varepsilon_0 hc)$ and that in the transition frequencies between states of different principle quantum numbers of hydrogen-like ions $M^{(Z-1)+}$ comprising a nucleus with $Z+$ electric charge and one electron. The transition frequency is approximately given by $f(M^{(Z-1)+}) = p\alpha Z^2 + q\alpha^2 Z^4$. The second term is negligibly small for $f(H)$. The ratio between the transition frequencies of an $M^{(Z-1)+}$ ion and an H atom is given by $Z^2 + (q\alpha^2 Z^4/p)$, for which the second term is non-negligible

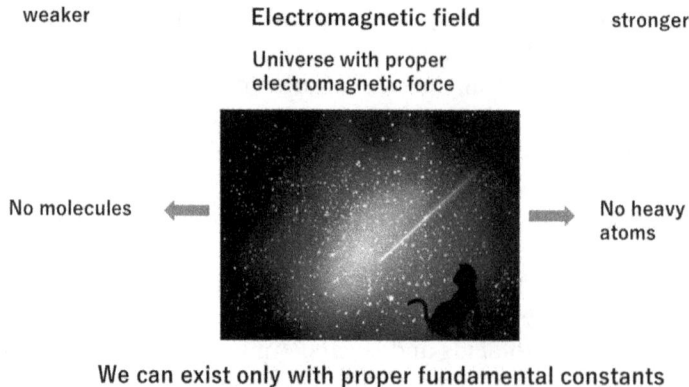

weaker   Electromagnetic field   stronger

Universe with proper electromagnetic force

No molecules ← → No heavy atoms

We can exist only with proper fundamental constants

Figure 6.14. The Universe would be quite different with stronger or weaker electromagnetic force.

for large Z. The variation in α is observed as the variation in the ratio between transition frequencies of an $M^{(Z-1)+}$ ion and an H atom. The transition frequencies of heavy atoms depend quite heavily on α because of the significant relativistic effect with the electron motion. Measurements in laboratories were also performed by comparing various atomic transition frequencies (table 6.1).

If there is a variation in the fine structure constant α, there should also be a variation in the proton-to-electron mass ratio μ ($= m_p/m_e$) [32]. The proton mass m_p is 45 times larger than the total mass of the constituent particles (two up-quarks and one down-quark). Therefore, the proton mass is dominated mainly by the binding energy, which changes with the variation in the electromagnetic force between quarks ($\propto\alpha$). For the search of the variation in μ, precise measurement of transition frequencies given by the nuclear motion is required. Atomic transition frequencies in the optical region are given only by the electron motion; therefore, the sensitivity to μ is very low.

Here, we consider the energy structures of molecules, having vibrational and rotational structures (figure 6.15).

The transition frequencies changing the vibrational states and rotational states are approximately proportional to $\mu^{-1/2}$ and μ^{-1}, respectively. Therefore, the precise measurement of the ratio of the vibrational transition frequencies to the atomic transition frequencies is useful in the search for variations in μ. Measurement of the molecular transition frequencies with uncertainty lower than 10^{-15} has never been reported. However, several proposals have been made for the measurement of vibrational transition frequencies with uncertainty lower than 10^{-16} [33–35]. In the near future, the search for variations in μ will be possible with uncertainty on the order of 10^{-16} yr^{-1}.

The measurement uncertainties of transition frequencies are still not low enough for the search for variations in fundamental constants. Observation of variations in several fundamental constants can give us important information for new physics, including the existence of extra dimensions. If there is temporal variation in the fundamental constants, the physical laws might have been different at the birth of the Universe.

6.9 Precise measurement in astronomical research

The fundamental activity of astronomy is the observation of the positions and motions of stars. As depicted in figure 6.16, the positions of stars are determined by

Table 6.1. Measurement results of the variation in fine structure constant α in laboratories.

Transitions	$(d\alpha/dt)/\alpha$ (yr^{-1})
^{199}Hg$^+$ $^2S_{1/2} - ^2D_{5/2}$ and ^{27}Al$^+$ $^1S_0 - ^3P_0$	$(-1.6 \pm 2.3) \times 10^{-17}$ [28]
^{171}Yb$^+$ $^2S_{1/2} - ^2F_{7/2}$ and ^{171}Yb$^+$ $^2S_{1/2} - ^2D_{3/2}$	$(-2.0 \pm 2.0) \times 10^{-17}$ [29]
	$(-0.5 \pm 1.6) \times 10^{-17}$ [30]
^{171}Yb$^+$ $^2S_{1/2} - ^2F_{7/2}$ and ^{87}Sr $^1S_0 - ^3P_0$	$(-4.3 \pm 2.5) \times 10^{-18}$ [31]

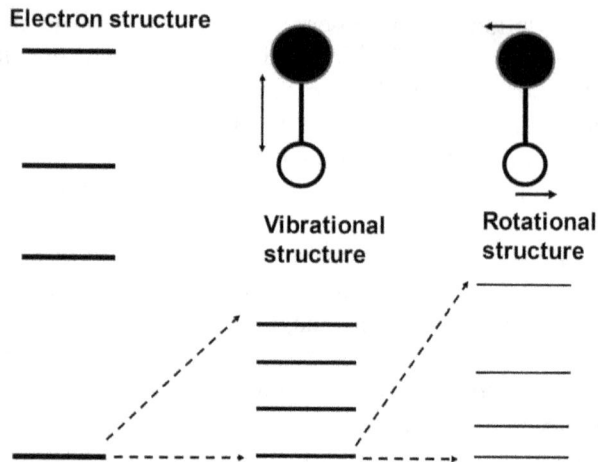

Figure 6.15. Schematic of the energy structure of diatomic molecules having vibrational rotational states.

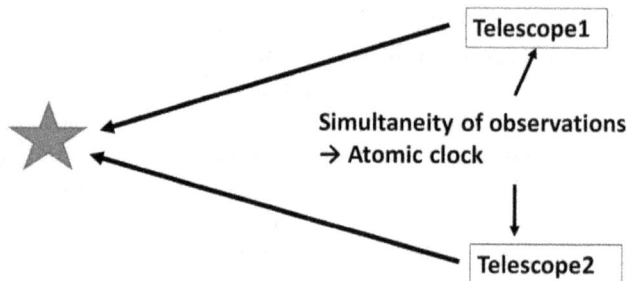

Figure 6.16. Measurement of the position of a star using two telescopes.

the displacement seen when they are observed from two or more telescopes at distant places, much like the binocular vision of humans and animals. The uncertainty of the direction gives the uncertainty of the positions of stars. Note also that the direction of observation changes as time elapses; therefore, simultaneity of observation is a very important factor in providing measurement uncertainty. Atomic clocks play a very important role in guaranteeing simultaneity with high accuracy.

Analyzing the radiation emanating from stars, we see there are particularly strong or weak intensities at specific frequencies (figure 6.17). Thus, we can discover the material compositions of celestial objects. The transition frequencies observed from stars are shifted because of the Doppler effect. From the shifts in the transition frequencies, we can determine the velocity of motion of the stars. The accuracy obtained for the shifts in the observed transition frequencies determines the accuracy of the information regarding the motion of stars.

From the precise measurement of the positions and velocities of stars, we can learn the history of the Universe since its birth with the Big Bang 13.8 billion years ago (figure 6.18). It is known that the Universe continued to expand and became transparent radiation 380 million years after its birth. The expansion of the Universe

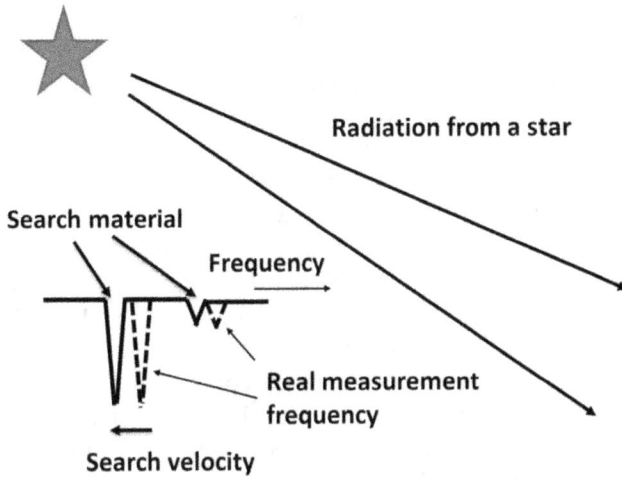

Figure 6.17. Measurement of the material and velocity of a star from the frequency distribution of radiation.

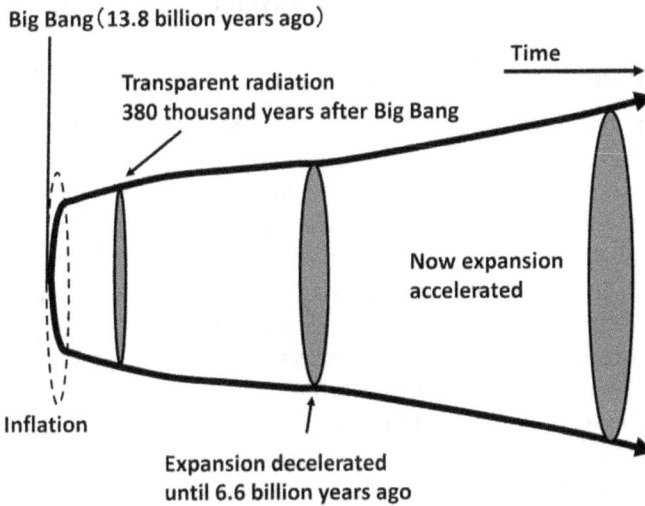

Figure 6.18. History of the Universe after the Big Bang.

has been accelerating since 6.6 billion years ago. This is the mystery called 'dark energy', which will be solved through further observation of the stars with low uncertainties.

6.10 Decay of protons

A grand unified theory (GUT) is a model in particle physics in which the strong nuclear force, electromagnetic force, and weak nuclear force are merged into one single force at high energy. If the GUT is realized in nature, there is a possibility that the Universe was very high energy (10^{27} K) so that the fundamental forces were unified.

Outside the nucleus, free neutrons are unstable with a mean lifetime of 881 s and decay to proton + electron + neutrino. GUT predicts the possibility of decay of a proton to neutral pion + positron or positive charged pion + neutrino. The mean lifetime of a proton is estimated to be 10^{31}–10^{36} years.

An experiment to detect proton decay was performed at Super-Kamiokande, having 500 million litres of distilled water at 1000 m underground in Kamioka, Gifu Prefecture in Japan. There, the cosmic ray signal giving a noise is four orders smaller than on the Earth. The decay of a proton from one of 10^{33} water molecules might be observed per several years. The lower limit of the mean lifetime was estimated to be 10^{34} years from antimuon decay [36]. After the construction of Hyper-Kamiokande (one billion litres of distilled water) is completed, a five- to ten-fold improvement in sensitivity is expected. Proton decay is caused when quarks in a proton get closer than 10^{-31} m (16 orders smaller than proton size). If the quarks are larger than 10^{-31} m, proton decay is not possible.

Proton decay should be observed with a signal from one water molecule, and all kinds of measurement noise must be suppressed for this purpose.

6.11 Precise measurement of time and frequency for physics

In this chapter, the role of the precise measurement of time and frequency in the development of physics was summarized. In ancient times, the precise measurement of time and frequency was much more difficult than that of length and mass. When new clocks were invented, there was significant progress in physics. Now, time and frequency are physical values that we can measure with the lowest uncertainty. In modern physics, the information regarding mass and length are obtained from the measurement of time and frequency. The development of modern physics has been made possible through the discovery of very small effects, which could be observed after the reduction of measurement uncertainties.

There are also examples to confirm the physical concepts, which seem to have been well established with the current measurement uncertainties (for examples, the constancy of the speed of light). This is because there is potential for new discrepancies with previous physics to be discovered after further reduction of measurement uncertainties.

References

[1] Filonovich S R 1986 *The Greatest Speed* (Moscow: Mir Publishers) p 285
[2] Fizeau H 1849 *C. R. Hebd. Seances Acad. Sci.* **29** 90
[3] Foucault L 1862 *C. R. Hebd. Seances Acad. Sci.* **55** 501
[4] Michelson A A 1881 *Am. J. Sci.* **22** 120
[5] Brillet A and Hall J L 1979 *Phys. Rev. Lett.* **42** 549
[6] Einstein A 1916 *Relativity: The Special and General Theory* translated 1920 (New York: H. Holt and Company)
[7] Mungall A G 1971 *Metrologia* **7** 49

[8] Chou C W 2010 *Science* **329** 1630

[9] Dyson F W *et al* 1920 *Philos. Trans. R. Soc.* A **220** 291

[10] Briatore L and Leschiutta S 1977 *Nuovo Cim.* B **37** 219

[11] Taylor J H and Weisberg J M 1982 *Astrophys. J.* **253** 908

[12] Abbott B P *et al* 2016 *Phys. Rev. Lett.* **116** 061102

[13] Atkins P W 1974 *Quanta: A Handbook of Concepts* p 52

[14] Christenson J H *et al* 1964 *Phys. Rev. Lett.* **13** 138

[15] Kobayashsi M and Maskawa T 1973 *Prog. Theor. Phys.* **49** 652

[16] Angelopoulos A *et al* 1998 *Phys. Lett.* B **444** 43

[17] Ulmer S *et al Nature* **524** 196

[18] Ahmadi M 2018 *Nature* **557** 71

[19] Andresen G B *et al* 2011 *Nat. Phys.* **7** 558

[20] Lamb W E and Retherford R C 1947 *Phys. Rev.* **72** 241

[21] Odom B *et al* 2006 *Phys. Rev. Lett.* **97** 030801

[22] Brodsky S J *et al* 2004 *Nucl. Phys.* B **703** 3

[23] Sick I and Trautmann D 2014 *Phys. Rev.* C **89** 012201

[24] Pohl R *et al* 2010 *Nature* **466** 213

[25] Quack M 2002 *Angew. Chem. Int. Ed.* **41** 4618

[26] Quack M 2008 *Annu. Rev. Phys. Chem.* **59** 741

[27] Dirac P A 1937 *Nature* **139** 323

[28] Rosenband T *et al* 2008 *Science* **319** 1808

[29] Huntemann N *et al* 2014 *Phys. Rev. Lett.* **113** 210802

[30] Godun R M *et al* 2014 *Phys. Rev. Lett.* **113** 210801

[31] Peik E 2017 *First North American Conf. on Trapped Ion*

[32] Calmet X and Fritzsch H 2002 *Eur. Phys. J.* D **24** 639

[33] Kajita M and Moriwaki Y 2009 *J. Phys. B: At. Mol. Opt.* **42** 154022

[34] Kajita M *et al* 2014 *Phys. Rev.* A **89** 032509

[35] Kajita M 2017 *Phys. Rev.* A **95** 023418

[36] Nishino H 2012 Super-K Collaboration *Phys. Rev. Lett.* **102** 141801

IOP Publishing

Measurement, Uncertainty and Lasers

Masatoshi Kajita

Chapter 7

Conclusion

This book has summarized the importance of recognizing non-zero measurement uncertainty, which can lead to significant problems in personal life or society. There are phenomena characterized by chaos, for which the solution is not predictable because a slight difference in initial condition leads to a significant difference in the solution.

The measurement of physical values was already done in ancient times, primarily length and mass, which are important for trade. At that time, different units were used in different regions. After the age of discovery, trade between different regions became active, and the use of different units between sellers and buyers led to difficulties.

After the French revolution, there was a movement towards the unification of the units used to measure physical values. Now units of seven physical values (mass, length, time, temperature, electric current, substance quantity, luminal intensity) are defined as SI units. With the reduction of measurement uncertainties, their definitions have been gradually changed.

After World War II, the invention of atomic clocks significantly reduced the measurement uncertainty of time and frequency. Moreover, the invention of the laser revolutionized physics so that the measurement uncertainties for all physical values could be reduced, via the definition of the speed of light, Planck constant, elementary electric charge, and Boltzmann constant. We can say that the definitions of physical values are changing from the macro scale to the micro scale because the properties of single atoms and molecules are universal.

It was also shown that the development of physics was closely correlated with the invention of new clocks. Now, physics researchers are showing intense interest in the precise measurement of time and frequency (also other physical values are measured from the time and frequency), which has very important roles in the search very small effects.

Also, researchers are trying to confirm established physical laws with reduced uncertainties, expecting the possibility of finding new discrepancies.

As an author, it would be a great pleasure if this book kindles an interest in young researchers regarding the importance of measurement uncertainties.

www.ingramcontent.com/pod-product-compliance
Lightning Source LLC
Chambersburg PA
CBHW082110210326
41599CB00033B/6657